SpringerBriefs in Immunology

For further volumes:
http://www.springer.com/series/10916

Awanish Kumar

Leishmania and Leishmaniasis

 Springer

Awanish Kumar
School of Life Sciences
Jawaharlal University
New Delhi, India

ISSN 2194-2773   ISSN 2194-2781 (electronic)
ISBN 978-1-4614-8868-2   ISBN 978-1-4614-8869-9 (eBook)
DOI 10.1007/978-1-4614-8869-9
Springer New York Heidelberg Dordrecht London

Library of Congress Control Number: 2013946651

Preface

This book provides substantial knowledge in sequence on the protozoan parasite *Leishmania* and disease globally caused by this parasite known as Leishmaniasis. The life cycle of leishmania is digenetic with vertebrates as definitive host and *Phlebotamine* sand fly as intermediate one. Over 350 million people are at risk of *Leishmania* infection, and at least 500,000 new cases of VL and 1.5 million cases of CL with severe morbidity are reported yearly. The rise of leishmaniasis is due to multiple factors including the AIDS epidemic, increase of international travel, lack of effective vaccines, difficulties in controlling vectors, international conflicts, and the development of resistance to chemotherapy. The book covers a range of *Leishmania*- and Leishmaniasis-specific information and disease pathogenesis. Introduction, immunology, invasion, clinical feature, drugs, vaccines, resistance, and experimental models of *Leishmania* are discussed in this book that aimed to gather broad knowledge for the students and researchers working in this field. The various matters discussed in this book collectively address the importance of this neglected disease. I would like to express deep sense of gratitude to my parents for their blessings inspiration and help in all stages of life. Finally, I thank SPRINGER, USA for their enormous efforts to publish this book. Lastly, I thank almighty, the Great, for giving me the strength, patience, and courage to carry out this uphill task.

New Delhi, India   Awanish Kumar

Abstract

Leishmaniasis is a poverty-associated disease with several different forms, like cutaneous leishmaniasis (CL), muco-cutaneous leishmaniasis (MCL), and visceral leishmaniasis (VL). CL has a spectrum of presentations, typically with self-healing or chronic lesions on the skin. MCL follows the complete resolution of the initial oriental sore; metastatic lesions appear on the buccal or nasal mucosa. VL is characterized by prolonged fever, splenomegaly, hepatomegaly, substantial weight loss, progressive anemia, pancytopenia, and hyperglobulimia. It is the most dreaded and devastating amongst the various forms of leishmaniasis and fatal without treatment. More than 350 million people are at risk of *Leishmania* infection, and at least 500,000 new cases of VL and 1.5 million cases of CL with severe morbidity are reported yearly. The rise of leishmaniasis is due to multiple factors including the AIDS epidemic, increase of international travel, lack of effective vaccines, difficulties in controlling vectors, international conflicts, and the development of resistance to chemotherapy. The book covers a range of *Leishmania*- and Leishmaniasis-specific information and disease pathogenesis. Introduction, immunology, invasion, clinical feature, drugs, vaccines, resistance, and experimental models of *Leishmania* are discussed in this book that aimed to gather broad knowledge for the students and researchers working in this field. The various matter discussed in this book collectively address the importance of this neglected disease. Scientists both from academic fields and from the industry involved in *Leishmania* research will find in this book a valuable and fundamental guide that conveys the knowledge needed to understand and to improve the success in combating this disease worldwide.

Keywords Leishmaniasis • Leishmania • Parasite • Life cycle • Vector • Host • Evasion • Survival • Disease • Symptoms • Infection • HIV • Immunology • Antigen • Adjuvants • Macrophage • Diagnosis • Drug • Antimonials • Resistance • Mechanism • Experimental models • Hamster • Vaccine

Contents

Chapter 1
Introduction

1.1 Historical Background of *Leishmania* and Leishmaniasis

Leishmaniasis caused by protozoan parasites *Leishmania*, is a disease of poverty as its victims are among the poorest. According to ranking after malaria it is a second most prevalent parasitic disease. Leishmaniasis has been considered a tropical affliction that constitutes one of the six entities on the list of most important diseases of World Health Organization/Tropical Disease Research (WHO/TDR) viz. Malaria, Schistosomiasis, Filariasis, Chagas disease, African Trypanosomiasis, Leishmaniasis, Leprosy, Tuberculosis (Desjeux et al. 2001).

Representations of skin lesions and facial deformities have been found on pre-Inca pottery from Peru and Ecuador dating back to the first century AD. There are evidences that some forms of leishmaniasis prevailed as early as this period (http://www.who.int/en). There are detailed descriptions of oriental sore by Arab physicians including Avicenna in the tenth century, who described it as Balkh sore from northern Afghanistan, and there are later records from various places in the Middle East including Baghdad and Jericho; many of the conditions were given local names by which they are still known. In the Old World, Indian physicians applied the Sanskrit term kala-azar (meaning 'black fever') to an ancient disease later defined as visceral leishmaniasis (VL).

Kala-azar was first noticed in Jessore in India in 1824, when patients suffering from fevers that were thought to be due to malaria failed to respond to quinine. By 1862 the disease had spread to Burdwan, where it reached epidemic proportions. In 1901, William Leishman identified certain organisms in smears taken from the spleen of a patient who had died from 'dum-dum fever'. Initially, these organisms were considered to be trypanosomes, but in 1903 Captain Donovan described them as being new. The link between these organisms and KA was eventually discovered by Major Ross, who named them *Leishmania donovani*. The search for a vector was a long one, and it was not until 1921 that the experimental proof of transmission to

A. Kumar, *Leishmania and Leishmaniasis*, SpringerBriefs in Immunology 3,
DOI 10.1007/978-1-4614-8869-9_1, © Springer Science+Business Media New York 2013

humans by sandflies belonging to the genus *Phlebotomus* was demonstrated by Edouard and Etienne. Swaminath et al. (2006) proved using human volunteers that the *Leishmania* parasite could be transmitted by *Phlebotomus* sandflies. The difficulties linked to vector (sandfly) control and the lack of an effective vaccine, the control of leishmaniasis relies mostly on chemotherapy.

1.2 Risk Factors and Definition of the Problem

In India, a country with a high leishmaniasis burden, 88 % of leishmaniasis patients have a daily income of less than US$2 and poor economic level (Murray 2005). The number of cases of leishmaniasis is increasing, mainly because of man-made environmental changes that increase human exposure to the sandfly vector (Desjeux et al. 2001). Extracting timber, mining, building dams, creating new irrigation schemes, expanding road construction, continuing widespread migration from rural to urban areas, and continuing fast urbanization worldwide are among the primary causes for increased exposure to the sandfly (Desjeux 2004).

1.3 Types of Leishmaniasis

The leishmaniasis causes considerable morbidity and mortality. It is the collective name for a number of diseases which have diverse clinical manifestations. Leishmaniasis has traditionally been classified in three major forms on the basis of clinical symptoms (Handman 2001). The most deadly form is visceral leishmaniasis (VL), which if left untreated, leads to death. A number of other species of *Leishmania* cause cutaneous (CL) and mucocutaneous (MCL) leishmaniasis, which, if not fatal, are still responsible for considerable morbidity of a vast number of people in endemic foci (Peters et al. 1983; Prasad 1999).

1.3.1 Cutaneous Leishmaniasis (CL)

This is the most common form of Leishmaniasis, also known as 'Oriental sore' which first appears as a persistent insect bite. Simple skin lesions appear at the site of sandfly bite (Fig. 1.1) which self-heal within few months but leaves scars. The incubation period can last from few days to months. Gradually, the lesion enlarges, remaining red. eat or pain. Resolution of the lesion involves immigration of leucocytes, which isolate the infected area leading to necrosis of infected tissues, and formation of a healing granuloma. CL is usually caused by *L. major, L. tropica, L. aethiopica,* in the old world and *by L. mexicana, L venezuelensis, L. amazonensis, L. braziliensis, L. panamensis, L. guyanensis* and *L. peruviana* in new world.

Fig. 1.1 Skin ulcer

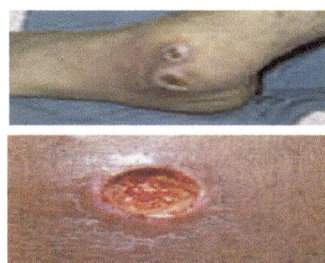

Fig. 1.2 DCL

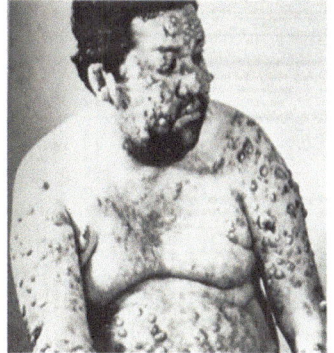

1.3.2   Diffuse Cutaneous Leishmaniasis (DCL)

This is a chronic, progressive, polyparasitic variant that develops in context of leishmanial-specific anergy and is manifested by disseminated non-ulcerative skin lesions, which can resemble lesions of lepromatous leprosy (Fig. 1.2). DCL is restricted to Venezuela and Dominican Republic in the western hemisphere, and to Ethiopia and Kenya in Africa. Its main causative organisms are *L. aethiopica* (old world) *and L mexicana* species complex (new world).

1.3.3   Mucocutaneous Leishmaniasis (MCL)

This form of disease, also known as "espundia", causes extensive destruction of naso-oral and pharyngeal cavities with hideous disfiguring lesions, mutilation of the face (Fig. 1.3) and great suffering for life. MCL is occasionally reported from Sudan and other Old World foci. Classical MCL is, however, restricted to *L. braziliensis* infections in which, following the apparently complete resolution of the initial oriental sore, sometimes many years later, metastatic lesions appear on the buccal or nasal mucosa. MCL usually exists as an azoonotic infection in which lifecycle is being transmitted from rodent to rodent and mammal by the forest sandfly *Lutzomyia spp.*

Fig. 1.3 MCL in the patient
with a perforated nasal
septum and mucosal tissue
destruction (Reproduced
from TDR Reports)

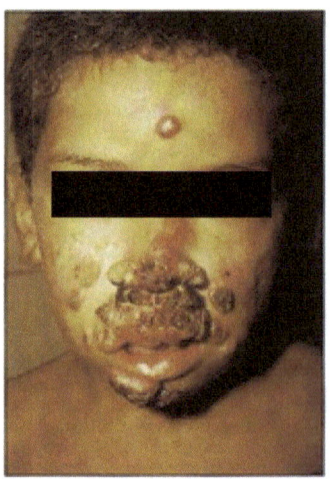

The reservoir hosts include rodents, opossums, anteaters, sloths and dogs etc.
The causative agents of MCL in old world are *L. aethiopica* (rare), and in new world
are *L. braziliensis, L. guyanensis, L. mexicana, L. amazonensis* and *L. panamensis.*

1.3.4   Visceral Leishmaniasis (VL) or Kala-azar (KA)

VL is the most dreaded and devastating amongst the various forms of leishmaniasis.
VL is also known as Kala-Azar, Black Sickness, Black Fever, Burdwan fever,
Dumdum fever or Sarkari Bimari etc. It is the most severe form of disease and if left
untreated, is usually fatal. The parasite is responsible for a spectrum of clinical syn-
dromes, which can, in most extreme cases, move from an asymptomatic infection to
a fatal form of VL. It is characterized by prolonged fever, splenomegaly, hepatomegaly, substantial weight loss, progressive anemia, pancytopenia, and hyperglobulimia. VL is complicated by secondary opportunistic infections (Fig. 1.4). The
parasite invades and multiplies within macrophages (free mononuclear phagocytic
cells) and affects the reticuloendothelial system including spleen, liver, bone marrow, and lymphoid tissue (Aggarwal et al. 1999; Boelaert et al. 2000). The outcome
of fully developed VL is death, usually said to be due to concomitant infection
resulting from the weakened immunological state of the patient. VL is typically
caused by *L. donovani* complex, which includes three species: *L. donovani donovani, L d. infantum,* and *L. d. chagasi. L. donovani* is the causative in the Indian
subcontinent and East Africa. *L. infantum* causes VL in the Mediterranean basin
and *L. chagasi* is responsible for the disease in Central and South America (Aggarwal
et al. 1999; Garg and Dube 2006; Singh 2006). VL is emerging as an important
opportunistic infection among people with HIV-1 infection (Alvar et al. 1997;
Desjeux and Alvar 2003). In fact, the parasite may be a cofactor in the pathogenesis
of HIV infection (Bernier et al. 1998). There are more than 21 morphologically
indistinguishable species of *Leishmania* that infect humans. Conventionally, they

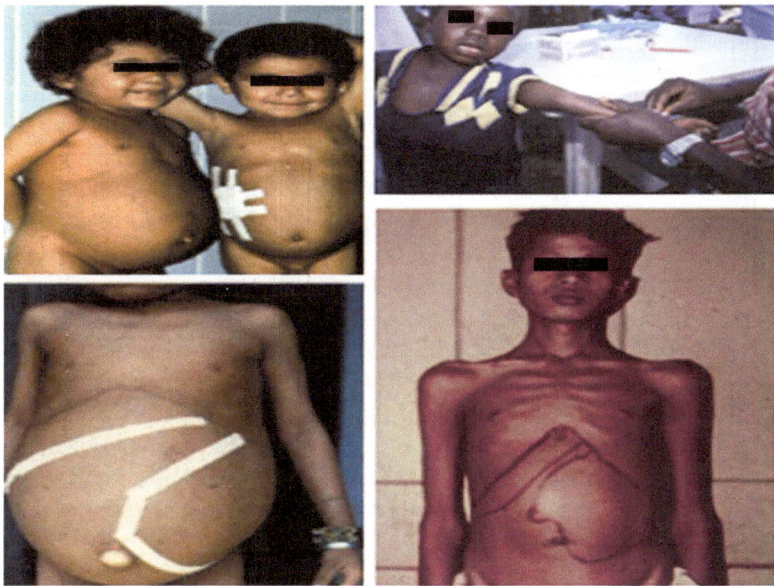

Fig. 1.4   Clinical symptoms of VL. Hepato-splenomegaly and substantial weight loss are main features (Reproduced from TDR Reports)

Fig. 1.5   PKDL patient with popular nodular lesions over face (Reproduced from TDR Reports)

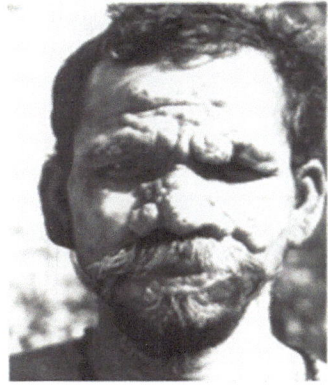

are classified and named mainly according to their geographical distribution and clinical characteristics of the disease they afflict (Bogitsh et al. 1999; Chang and Chang 1985; Herwaldt 1999; Roberts et al. 1996).

The **Post kala-azar Dermal Leishmaniasis (PKDL)** is a type of non ulcerative cutaneous lesion (Fig. 1.5). After recovery from infection, VL patients may develop a chronic form of CL i.e., PKDL which is developed in about 10 % of kala-azar patients generally 1 or 2 years after completion of sodium antimony gluconate (SAG) treatment and requires a long and expensive treatment (Rees et al. 1984; Salotra and Singh 2006).

1.4   Geographical Distribution of Leishmaniasis

Leishmaniasis occurs in 88 countries in tropical and temperate regions, of which 72
are either developing or least developed. Approximately 1,98,000 people are affected
with these diseases worldwide with 5,00,000 million new cases occurring each year
(WHO 2001) but the true picture remains largely hidden since a substantial number
of cases are never recorded (Herwaldt 1999). The disability-adjusted life years
(DALY) burden was 2,357,000 and total deaths were 59,000 in 2001 (WHO Health
Report 2002). It has been estimated that 90 % of CL cases occur in 7 countries:
Afghanistan, Algeria, Brazil, Iran, Peru, Saudi Arabia and Syria whereas MCL is
endemic in Mexico, Central and South America (Fig. 1.6a). Annual estimate for the
prevalence of kala-azar cases worldwide is 0.5 million and 2.5 million, respectively
(Croft et al. 2006; Desjeux 2004) and of these, 90 % cases occur in India, Nepal,
Bangladesh and Sudan. In India VL is most prevalent in Bihar, West Bengal, Assam
and Eastern Uttar Pradesh. A summary of clinical manifestations and geographic
distribution of the *Leishmania* species is summarised in Tables 1.1 and 1.2.

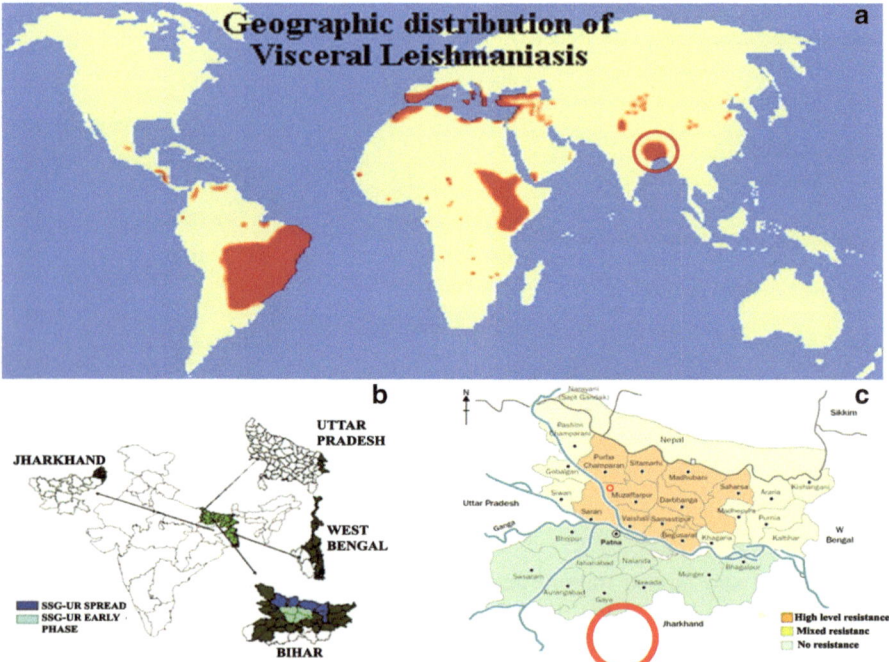

Fig. 1.6 Geographical distribution of visceral leishmaniasis (VL); (**a**) Worldwide distribution of
VL, (**b**) VL affected states of India and (**c**) VL affected districts of Bihar

Table 1.1 Leishmania species cause human disease, geographic location, reservoir animals and vectors

Organisms	Infection	Geography	Reservoir	Vector
L. donovani	Visceral leishmaniasis	North-east India, Bangladesh, Burma	Humans	*Phiebotomus argentipes*
L. infantum	Visceral leishmaniasis	Mediterranean basin, Middle east, China, Asia	Dogs, foxes, jackals	*P. pcrniciosufi, P. arias*
L. donovani	Visceral leishmaniasis	Sudan, Kenya, Horn of Africa	Rodents in Sudan, Canines, Humans	*P. orinntalis, P. martini*
L. major	Cutaneous leishmaniasis	Semideserts in Middle East, North India, Pakistan	Gerbils	*P. papatassi*
L. major	Cutaneous leishmaniasis	Sub-Sahara, Sudan	Rodents	*P. duboscqi*
L. tropica	Cutaneous leishmaniasis	Middle East, Mediterranean basin, central Asia	Humans	*P. sergenti*
L. aethiopica	Cutaneous leishmaniasis	Highlands of Kenya, Ethiopia	Hyraxes	*P. longipes, P. pedifer*
L. chagasi	Visceral leishmaniasis	Central, Northern South America, Brazil, Venezuela	Foxes, dogs, opossums	*Luizomyia longipalpis*
L, mexicana	Cutaneous leishmaniasis	Yucatan, Guatemala	Forest rodents	*L. olmeca*
L. amazonensis	Cutaneous leishmaniasis	Tropical forests of South America	Forest rodents	*L. flaviscutellata*
L. braziliensis	Mucocutaneous leishmaniasis	Tropical forests of South and Central America	Forest rodents, peridomestic animals	*Psychodopygus Lutzomyia spp.,*
L.. guyanemis	Mucocutaneous leishmaniasis	Guyana, Surinam	Sloths arboreal anteaters	*L. umbratilis*
L.. panamensis	Mucocutaneous leishmaniasis	Panama, Costa Rica, Colombia	Sloths	*L. trapidoictal*
L. peruviana	Mucocutaneous leishmaniasis	West Andes of Peru. Argentine highlands	Dogs	*L. verrucarurn, L. pvmenis*

Table 1.2 Summary of clinical manifestations and geographical distribution of *Leishmania* species

Species	Clinical manifestation	Geographical distribution
L. donovani	Visceral (kala-azar)	Old World: China, India, Bangladesh
L. infantum	Visceral	Old World: North Central Asia, Northwest China, Uzbekistan, Middle East
L. chagasi	Visceral	New World: South and Central America
L. major	Cutaneous	Old World: Africa, Middle East, Northern Asia
L. mexicana	Cutaneous, Diffuse cutaneous, Mucocutaneous	New World: Southern Mexico, Belize, Northern Guatemala, Southern Texas
L. amazonensis	Cutaneous, Mucocutaneous	New World: South and Central America
L. braziliensis	Mucocutaneous, Cutaneous	New World: Throughout South America

1.5   Global Status of Visceral Leishmaniasis (Kala-azar)

VL is endemic in 62 countries, with 200 million people at risk, an estimated 500,000 new cases each year worldwide (Herwaldt 1999; Guerin et al. 2002; Desjeux 2004; Thakur 2003) and 41,000 recorded deaths in the year 2000 (WHO 2001). The disease burden associated with VL, measured in DALYs was estimated to be 1,980,000 (1,067,000 for male and 744,000 for female (Guerin et al. 2002) in year 2000. VL is caused by *L. donovani* in the Indian subcontinent, Asia, Africa and by *L. infantum* in the Mediterranean region, southwest and central Asia and by *L. chagasi* in South America. In Sudan, for example, a major decade-long epidemic of VL occurred from 1984 to 1994. As this was the first epidemic in the area, populations were highly susceptible. Some studies estimate that the disease caused 100,000 deaths in a population of around 300,000 in the western upper Nile area of the country (http://www.who.int/mc/diseases/leish/diseaseinfo.htm). The health ministers of India, Nepal and Bangladesh signed a memorandum of understanding on 18 May 2005, pledging to eliminate VL from their countries. The five elements of the elimination strategy are access to early diagnosis and treatment, strengthening disease and vector surveillance, integrated vector management, social mobilization, networking and research (WHO/TDR News 2005).

VL is present in India for more than 100 years. The incidence of KA in India is among the highest in the world (Desjeux 1992). Epidemics of KA occurred in Bengal in the years 1832, 1857, 1871, 1877, and 1899. In India about 100,000 cases of VL are estimated to occur annually (TDR News No. 37, Nov, 1991). It has recently posed a serious threat in India involving 38 out of 42 districts of Bihar state, 8 districts of West Bengal and 2 districts of Eastern Uttar Pradesh (Bora 1999) (Fig. 1.6b,c). In 1977, a sample survey in Bihar estimates the number of cases to be about 1,00,000 with 4,500 deaths whereas in 1991 infected cases reached to 2,50,000 with 75,000 deaths (Modabber 1990, 1995; Thakur et al. 1993). The state of Bihar now accounts for more than 90 % of cases (Zijlstra et al. 1995). Because of the rapid manner in which the disease was spreading, an alarming situation existed (Desjeux 1992).

1.6   Morphology and Life Cycle of *Leishmania donovani*

In India, VL is caused by *L. donovani*. Indian VL is anthroponotic and is transmitted chiefly through the bites of the female sandfly, *P. argentipes*. *Leishmania* exists in two forms (i) promastigotes: these are extracellular, elongated, flagellated, motile and ranges in size from 2 μm × 2–20 μm (Fig. 1.7a). This form exists in sandfly and in *in vitro* cultures (ii) amastigotes: these are intracellular, round to oval, aflagellated, non-motile and ranges in size from 2 to 5 μm (Fig. 1.7b). This form resides and multiplies within the phagolysosomes of macrophages of reticuloendothelial system of the vertebrate host (Handman 1999).

Following the sandfly bite, some of the flagellates entering the circulation are destroyed while others enter the cells of the reticuloendothelial system. Here they undergo change into amastigote form which multiplies by binary fission, with the

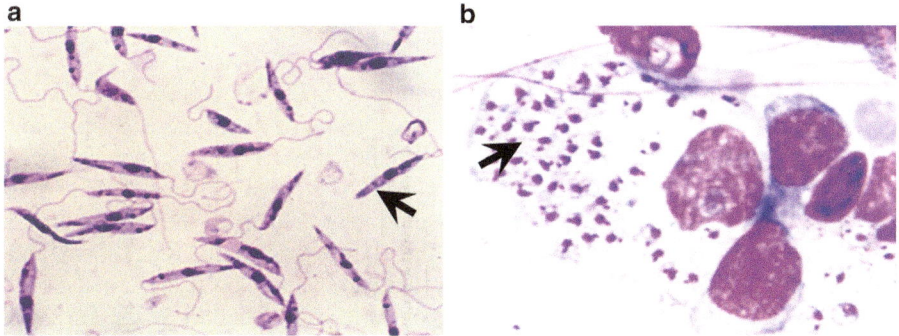

Fig. 1.7 Two stages of *Leishmania* parasite; (**a**) Extracellular and motile form called promastigotes each bearing a flagellum. (**b**) Intracellular and non-motile stage called amastigotes (*small dots*) as seen in Giemsa stained dab smear prepared from the spleen of *L. donovani* infected golden hamster

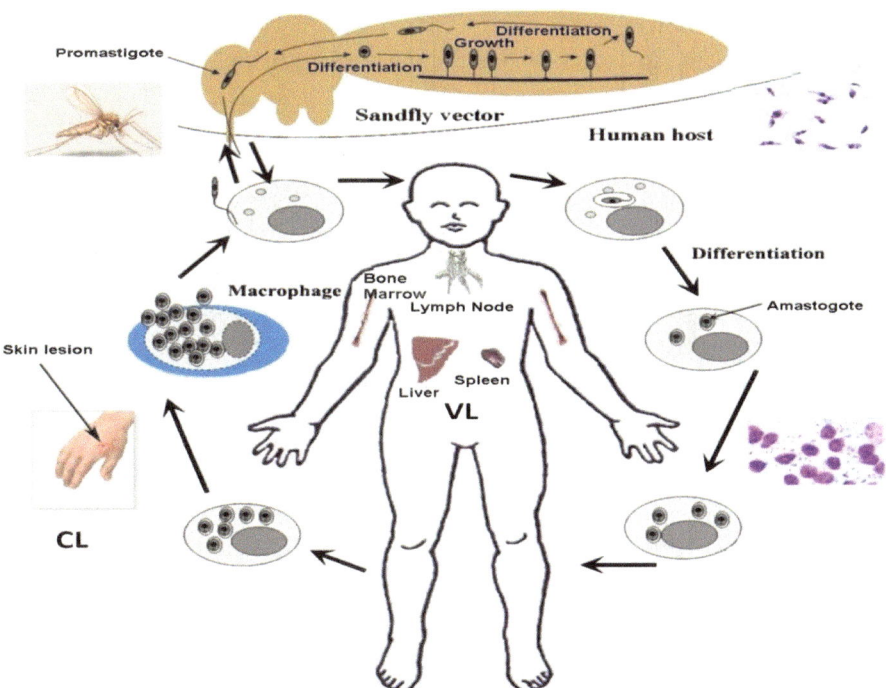

Fig. 1.8 Life cycle of *Leishmania* parasite shuttles between vector sandfly and human host

multiplication continuing until the host cell is packed with the parasites and ruptures, liberating the amastigotes into circulation. The released amastigotes are taken up by additional macrophages and so the cycle continues (Fig. 1.8). Ultimately all the organs containing macrophages and phagocytes are infected, especially the spleen, liver and bone marrow.

1.7   Vectors and Transmission of the Disease

Leishmaniasis is transmitted by the *Phlebotomus spp.* in the old world and *Lutzomyia spp.* in the new world. *P. argentipes* is the proved vector of KA in India (Swaminath et al. 2006; Kishore et al. 2006). Of the 500 known phlebotomine species, only some 30 of them have been positively identified as vectors of the disease. Shadflies are very small in size (<3.5 mm) (Fig. 1.9) and may be hard to see. Sandflies are usually most active in the twilight, evening and night hours (from dusk to dawn) and less active during the hottest time of the day. Female shadfly lays its eggs in the burrows of certain rodents, in the bark of old trees, in ruined buildings, in cracks in house walls, in animal shelters and in household rubbish (http://www.who.int/en). High incidence of VL is reported during pre-monsoon season that coincides with vector abundance and increased man-vector contact due to sleeping habits of children in open space (Singh 2006; Kishore et al. 2006).

1.8   Clinical Symptoms of Visceral Leishmaniasis

VL patients are heavily infected throughout the mononuclear phagocyte system; develop life-threatening disease after an incubation period of weeks to months; and have fever, severe cachexia, hepatosplenomegaly (Fig. 1.10), pancytopenia

Fig. 1.9 Sandfly, the vector host of *Leishmania* parasite (Reproduced from TDR Reports)

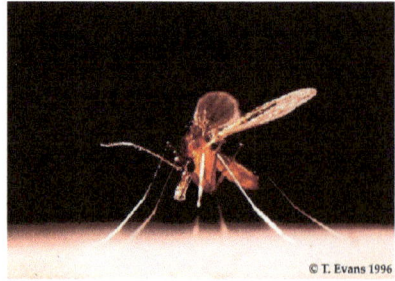

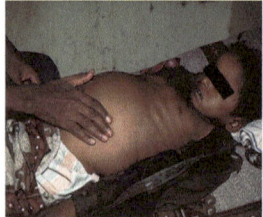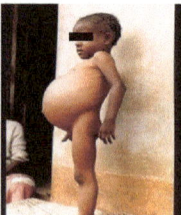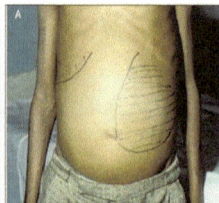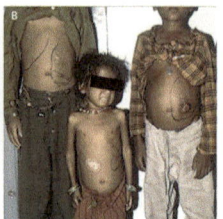

Fig. 1.10 Clinical symptoms of VL; Hepato-splenomegaly and wasting are the main features (Reproduced from Murray et al., (2005); Lancet 366: 1561–77 and TDR reports)

(anaemia, thrombocytopenia, and leucopenia, with neutropenia, marked eosinopenia, and a relative lymphocytosis and monocytosis), and hypergammaglobulinaemia (mainly IgG from polyclonal B-cell activation) with hypoalbuminaemia (Herwaldt 1999). VL encompasses a broad range of manifestations of infection which remains asymptomatic or subclinical in many cases, or can follow an acute, subacute, or chronic course. Active VL may also represent relapse (recurrence 6–12 months after successful treatment) or late reactivation of subclinical or previously treated infection.

1.9   Control Strategies of the Disease

Efficient case management based on early diagnosis and treatment is the key to limit morbidity and prevent mortality. Effective treatment of patients is also a measure to control reservoir and transmission in anthroponotic foci, particularly for cases of PKDL, which are thought to act as a long term reservoir of the disease. In addition, vector control should be implemented wherever feasible. Spraying of houses with residual insecticides has been an important measure in the past in India but is not much used now. Insecticides used in malaria control programmes are effective on sandfly. In foci where sandflies bite at night, impregnated bednets have decreased the incidence of leishmaniasis. Indian government started Leishmaniasis Elimination Programme in 2001 with the targets of prevention of death by 2004, zero level incidence by 2007, zero level prevalence by 2010 and elimination by 2012 (Ganguly et al. 2006).

1.10   Leishmania/HIV Co-infections

Leishmania/HIV co-infection is emerging as an extremely serious problem. To date, it has been reported from 34 countries, with most of the cases from Southern Europe, where 25–70 % of adult patients with VL are co-infected with HIV (Desjeux et al. 2001). The first case of leishmaniasis associated with human immunodeficiency virus (HIV) infection was published in middle of the 1980s. The first case of VL/HIV co-infection in India was identified from the State of Bihar in the year 2000 (Sinha et al. 2002). Since AIDS epidemic is looming large on the horizon of new millennium in India (Mukhopadhyay et al. 1996; Sinha et al. 2006), the state of Bihar needs to be looked seriously for VL/HIV co-infections. The size of CD4+ cells count is $\leq 200/\mu m^3$ in 62–90 % of the co-infected patients. The majority of the cases of Leishmania/HIV co-infection have been described in adults infected by HIV-1; however, they have also been reported in patients infected by HIV-2 (Alvar et al. 1997). These co-infections impose specific difficulties in terms of diagnosis and treatment (often results in frequent failure and relapses due to drug resistance). The development of the HIV/AIDS pandemic during the last 20 years has modified the spectrum of leishmaniasis in both the clinical and epidemiological fields.

Chapter 2
Immunology of *Leishmania*

2.1 Introduction

In view of existing knowledge, lymphocytes, macrophages (MΦ) and antibody co-operation the major components of the immune system are necessary for protection against the leishmania infection. The death of the parasite occurs either by stimulation of the sensitized lymphocytes through antigen load or by activating macrophages. Active VL disease is characterized by the marked elevation of humoral immune response i.e. by the production of plenty of specific as well as non-specific antibodies (Ghose et al. 1980; Ghosh et al. 1995). Antibody titers, primarily IgG rise sharply during VL, but the antibodies so generated, are apparently not protective (Bryceson and Turk 1971; Bray and Lainson 1966; Evans et al. 1990). The enormous increase in serum immunoglobulin levels in active VL is due to poly-clonal activation of immunoglobulin producing cells leading to increase biosynthesis of IgG and to a lesser extent of IgM. Most of the antibodies produced during infection are not parasite specific (Clinton et al. 1969; Bunn-Moreno et al. 1985; Evans et al. 1990), but the hyper-gammaglobulinemia may have diagnostic value, IgG may reach 50 g/l and comprise 50 % of total serum proteins (Stauber 1963).

It is held that CMI response impairment causes the pronounced immunosuppression during the active stage of the disease. Rezai et al. (1978) noted a marked decrease in circulating T-lymphocytes in kala-azar patients. These observations help to explain the lack of delayed type of hypersensitivity (DTH) or protective cell-mediated immunity during VL. It has not yet been ascertained whether the lack of DTH to leishmania during infection is the result of general or specific immunosuppression phenomenon. After the successful treatment, CMI response tends to be restored and the test becomes positive (Sacks et al. 1987). Past studies explicit that in control and resolution of leishmaniasis, cell mediated immune response play a very important role in both experimental models of infection and human patients (Murray 1982; Howard et al. 1984; Stobie et al. 2000). The host's immunological responses during *L. major* infection has been studied thoroughly (Liew 1993; Locksley and Louis 1992). Resistance and susceptibility to *L. major* is mediated by

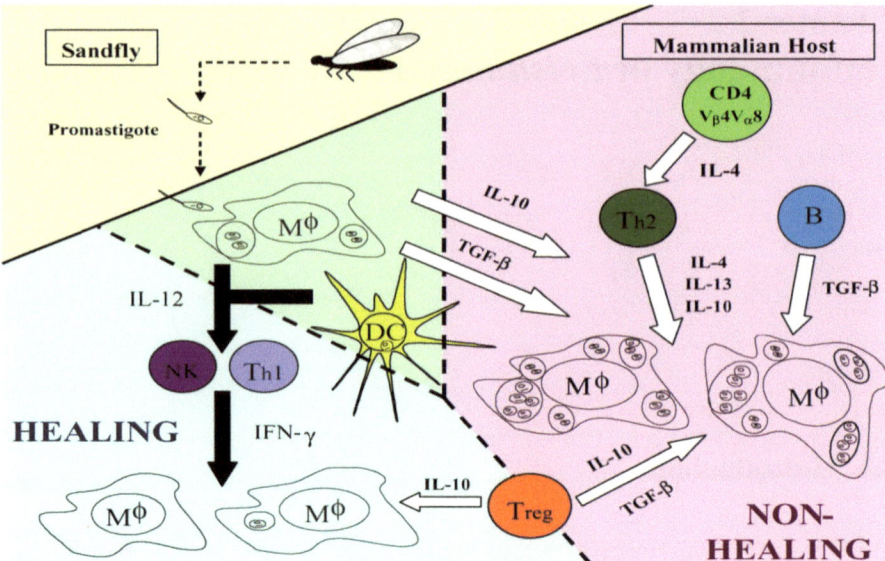

Fig. 2.1 Immunological responses in *Leishmania*: During a blood meal an infected sandfly trans-
mits metacyclic promastigotes to the vertebrate host, which convert to the amastigote form on
entering macrophages and dendritic cells. IL-12 production from infected cells induces NK cell
activation and CD4+ T helper-1 differentiation and IFN-γ production. IFN-γ stimulates iNOS
expression and NO production in the macrophage, which mediates parasite killing and therefore a
healing response. Failure to produce IL-12 or alternatively IL-4/IL-13 production results in unreg-
ulated parasite replication within the infected cells facilitated by host cell IL-10 production. IL-10
production by CD4+ T regulatory cells can both facilitate non-healing disease as well as maintaining latent infection and concomitant immunity (Diagram courtesy of James Alexander and Karen
Bryson, Immunology Letters (2005) 99; 17–23)

Th1 and Th2 subsets of CD4+ T cells, respectively. Th1 cells secrete protective
cytokine IFN-γ, IL-2 and execute CMI responses, where as Th2 cells produce disease exacerbative cytokines IL-4, IL-10 and IL-13 (Fig. 2.1) and assist in antibody
production for humoral immunity (Lehn et al. 1992; Heinzel et al. 1991; Holaday
et al. 1991; Suffia et al. 2000). Similarly Th1/Th2 dichotomy in cytokine response
to *L. donovani* infection in murine and human system has also been reported (Kemp
et al. 1993; Karp et al. 1993).

Protozoa of the genus *Leishmania* cause cutaneous, mucocutaneous or visceral
diseases in man depending on the species of the parasite and the host immune
response. While extensive information is available about the immune response in
experimental cutaneous leishmaniasis, the nature of immunity in experimental visceral leishmaniasis, which is different in many aspects, poorly understood. In order
to develop vaccines for different forms of leishmaniasis and since there are many
areas where different species and different forms of the disease overlap, a detailed
knowledge of the particularity of the immune response and pathogenesis are
extremely important. There are marked differences in immunity between
experimental visceral leishmaniasis and cutaneous leishmaniasis.

2.1.1 Role of B Cells and Immunoglobulin

Active kala-azar disease is characterized by the marked elevation of humoral immune response i.e. by the production of plenty of specific as well as non-specific antibodies (Ghose et al. 1980; Ghosh et al. 1995). Antibody titers, primarily IgG rise, sharply during visceral leishmaniasis, but the antibodies so generated, are apparently not protective (Bryceson and Turk 1971; Bray and Lainson 1966; Rezai et al. 1978; Evans et al. 1990). The enormous increase in the serum immunoglobulin levels in active kala-azar is due to polyclonal activation of immunoglobulin producing cells leading to increase biosynthesis of IgG and to a lesser extent of IgM. Most of the antibodies produced during infection are not parasite specific (Bunn-Moreno et al. 1985; Evans et al. 1990), but the hyper-gammaglobulinemia may have diagnostic value (Sen Gupta 1962), γ-globulin; may reach 50 g/l and may comprise 50 % of the total serum proteins (Stauber 1963). Manson-Bahr (1971) noted positive immediate skin reactions during active infection, but not delayed type of hypersensitivity.

Polyclonal B cell activation is present both in human and experimental visceral leishmaniasis (Bunn-Moreno et al. 1985), but the actual role of B cells or immunoglobulins in the immunity in visceral leishmaniasis have been poorly evaluated. Most of the data in this area have been obtained with cutaneous leishmaniasis models, where resistance was observed with depletion of B cells using anti-IgM antibody or in BALB xid mice, lacking B-1 B cells and susceptibility was increased by transfer of B cells or administration of B-cell hematopoietic factor, IL-7 (Sacks et al. 1987). In visceral leishmaniasis, enhanced resistance was recently shown in mutant mice that lack mature B cells. To distinguish the effect of B cells from that of immunoglobulin on susceptibility experiments were performed using mice genetically altered to contain no circulating antibody, with or without functional B cells, and mice defective in Fc receptor. These studies showed that the circulating antibody is crucial for susceptibility to the development of cutaneous leishmaniasis. Furthermore, amastigotes from the lesion of cutaneous leishmaniasis were shown to be coated by IgG, and internalization of Ig-coated amastigotes by macrophages was shown to lead to IL-10 production and consequent enhancement of intracellular parasite growth *in vitro*. Similar mechanisms might be acting in visceral leishmaniasis.

2.1.2 Role of T Lymphocytes

In human and experimental leishmaniasis immunity is predominantly mediated by T lymphocytes. Initial studies on mice using T cell-depleted mice and nude BALB/c mice (Murray 1982) have shown the importance of T lymphocytes for protection against *L. donovani* infection. Adoptive transfer of T cells, immune to *Leishmania* antigen, conferred resistance against *L. donovani* infection (Rezai et al. 1978). Reconstitution experiments using nude BALB/c mice, and cell depletion experiments in euthymic mice using monoclonal anti-CD4 or anti-CD8 antibodies showed the necessity of both CD4+ and CD8+ T cells in the protection against *L. donovani* infection (Sternberg et al. 1989). In *L. donovani*-infected BALB/c mice, there was a

time course-related participation of different cell populations: L3T4$^+$ (CD4$^+$) cells
are important in the initial two weeks of infection, when the parasite replication is
still occurring, mainly in the formation of hepatic granulomas, but later this cell
population decreases and is replaced by Lyt 2$^+$ (CD8$^+$) cells when the progress of the
infection is controlled. However, other studies point to the differences in the participation of immune elements in the protection observed in immune animals upon
re-infection compared with those in naive animals with a primary infection. In
immune BALB/c mice, depletion of Lyt 2$^+$ but not L3T4$^+$ abolishes resistance.
Furthermore, granuloma formation is not affected by depletion of Lyt2$^+$ or L3T4$^+$.
Conversely, cyclophosphamide-A treatment abolishes granuloma formation but
does not interfere with parasite replication (Murray et al. 1983).

2.1.3 Role of Cytokines

T lymphocytes participate in the immune response to *L. donovani* infection by producing different cytokines. While euthymic *L. donovani*-infected BALB/c mice are
able to control infection with granuloma formation and IFN-γ and IL-2 production,
nude BALB/c mice neither form granulomas nor produce IFN-γ (Murray 1982).
Human recombinant IFN-γ restores the ability of nude BALB/c mice to control
L. donovani infection. Furthermore, Anti-IFN-γ antibody abolishes granuloma formation (Squires et al. 1989), confirming the importance of this cytokine in protection. Moreover, depletion experiments using anti-IL-2 monoclonal antibodies and
reconstitution using recombinant IL-2 showed a role for IL-2 in leishmanicidal
activity apparently through the induction of IFN-γ, and in granuloma formation
with involvement of L3T4$^+$ and Lyt 2$^+$ T cell. Results in immune mice differ from
those described for naive animals. Neutralization of IL-2 but not IFN-γ abolishes
resistance, and granuloma formation is not affected by neutralization of IL-2 or
IFN-γ (Murray et al. 2005).

The role of IL-4 as a cytokine related to susceptibility has been recently ques-
tioned in experimental cutaneous leishmaniasis, but most studies on experimental
visceral leishmaniasis had raised this question about the role of IL-4 in susceptibility from the beginning. In one study, besides predominant IFN-γ production in the
initial and late phase of infection, IL-4 production was detected in the intermediary
phase coinciding with peak of parasite burden in the susceptible strain, and no IL-4
production in the resistant mouse strain (Saha et al. 2006). Nevertheless other studies contradicted these findings. No IL-4 or IL-5 production was observed in three
different strains of mice infected with *L. donovani*, and in the liver, only IFN-γ RNA
was detected by Northern blot, and both Th1 and Th2 cytokine mRNAs, IL-4,
IL-10, IFN-γ and IL-2 mRNA were detected by PCR. Furthermore, mice treated
with anti-IL-4 monoclonal antibodies and mice with IL-4 gene disruption did not
show better control of the infection. IL-10, another Th2 cytokine, however, was
related to progressive disease in human visceral leishmaniasis and was shown to
have a role in susceptibility in experimental visceral Leishmaniasis. A progressive

increase in IL-10 mRNA level in tissues during infection suggested a role in suscep-tibility. In addition, the control of parasite growth in inner organs in BALB/c IL-10-/- mice and in normal mice with IL-10 receptor blockade by antibodies confirmed the role of IL-10 in susceptibility (Murray et al. 2005). Since IL-10 receptor block-ade increased serum IFN-γ levels, a protective effect was initially attributed to the non-suppressed leishmanicidal effect of IFN-γ. However, suppression of parasite growth with IL-10 receptor blockade even in IFN-γ gene-disrupted mice suggested a broader effect of IL-10 on the suppression of multiple leishmanicidal mechanisms (Murray et al. 2005). We should emphasize, however, that visceral leishmaniasis in mice is a self-controlled infection; therefore IL-10 is probably not responsible for uncontrolled progressive increase in the parasite burden, which does not occur in this model. Its role should rather be considered similar to that in cutaneous leish-maniasis in resistant strain of mice, in which important findings have been recently obtained. In *L. major*-infected resistant C57BL/6 mice, IL-10 was shown to be important for the persistence of the parasite in the lesion, preventing its complete clearance from the lesion despite the presence of a protective immune response. Furthermore, this apparently undesirable persistence of the parasite was shown to be of the utmost importance for the maintenance of protective immunity against re-infection, with CD4$^+$CD25$^+$ regulatory T cells with IL-10-dependent and IL-10-independent mechanisms probably involving transforming growth factor TGF-β being involved in the suppression of IFN-γ production.

In contrast, IL-12 was shown to be linked to protection against the infection. IL-12 treatment of *L. donovani*-infected BALB/c mice significantly reduced the parasite burden with the participation of CD4$^+$ and CD8$^+$ T cells, NK cells and IFN-γ, IL-2 and tumor necrosis factor TNF-α. But a distinct antimicrobial effect of IL-12, independent of IFN-γ, was also demonstrated in experiments using IFN-γ gene-disrupted mice. These mice, as expected, show a progressive infection for the first eight weeks but in the late phase develop a capacity to reduce the parasite bur-den with the participation of TNF-α induced by IL-12. Neutralization of IL-12 with anti-IL-12 monoclonal antibody (Murray 1997) or IL-12 deficiency (IL-12-/-) shows the fundamental role of this cytokine in the control of infection in susceptible mice. Furthermore, increased levels of TGF-β were observed in IL-12-/- mice and were further increased in IL-12/IFN-γ double knockout mice, without expansion of the Th2 response.

2.1.4 Immunosuppression

One of the immunopathological consequences of active visceral leishmaniasis in humans is suppression of the T-cell responses mainly to *Leishmania* antigen. Although the *L. donovani*-infected mouse is not a good model for the study of immune suppression, negative *Leishmania* antigen-induced delayed-type hypersen-sitivity can be observed coinciding with the peak of parasite burden in the suscep-tible mouse strain and therefore some studies have been conducted using this model.

However, the better model to study this aspect is a hamster infected with *L. (L.) donovani* or *L. (L.) chagasi* that develop progressive visceral leishmaniasis. We have studied immunosuppression in *L. (L.) chagasi*-infected hamsters and have observed a concanavalin-A induced lymphoproliferative response in all experimental periods but the total absence of a *Leishmania* antigen induced response (Goto and Lindoso 2004). In the literature, the *Leishmania* antigen induced response was found to be suppressed in all studies, but there was disagreement about the concanavalin A-induced response. Some studies showed that the response was preserved during the experiment (Bunn-Moreno et al. 1985). Antigen-specific T-cell anergy present during active disease recovers after treatment and cure.

Various factors have been reported to cause immunosuppression in studies using either mouse or hamster models. Studies on mice have indicated T cells and others Th2 cells and adherent cells as being responsible for suppression. Macrophage mediated suppression is reported to lead to increased parasite growth and to be linked to either defective antigen presentation, suppression of class I or class II major histocompatibility complex molecule expression or mediation by prostaglandin-like substances (Murray et al. 1983; Reiner and Locksley 1987). In *L. (L.) donovani*-infected hamsters, adherent splenic cells have been shown to be important in the suppression of lymphoproliferation and in defective antigen presentation (Bogitsh et al. 1999). TGF-β produced by adherent antigen-presenting cells from infected hamsters was implicated in immunosuppression since a high level of TGF-β was observed in the cell culture supernatant when the *Leishmania* antigen-induced lymphoproliferative response was inhibited (Rodrigues et al. 1998). We have studied the effect of another growth factor, insulin-like growth factor-I, and we have shown its effect on *in vitro Leishmania* growth but also on enhancement of the lesion in cutaneous leishmaniasis (Goto and Lindoso 2004). More recent data suggest that it is a suppressor factor of macrophages leading to the decreased production of nitric oxide in *Leishmania*-infected macrophages *in vitro*. As mentioned in a previous section, cytokine IL-10 has been studied as a susceptibility factor in cutaneous leishmaniasis, but in our opinion it should also be considered within the context of immunosuppression. The absence of data in this field may be due to the difficulty to study hamsters, which is the most appropriate model, for which no reagents for cytokines are available. Only recently RT-PCR primers for some cytokines have been developed and using these primers, no qualitative change in the expression of different cytokine RNA was observed during visceral leishmaniasis in hamsters (Melby et al. 2001), suggesting the necessity to develop a more sensitive method for evaluation such as quantitative PCR.

On the basis of elements shown to participate in suppression, such as T cells and TGF-β and possibly IL-10, it becomes attractive to speculate on the possible participation of CD4[+]CD25[+] regulatory cells in immunosuppression during visceral leishmaniasis. Another aspect addressed in a number of studies of immunosuppression is the initial interaction between antigen-presenting cells and T cells. Decreased expression of costimulatory molecules B7-1 and Th1-specific M150 protein (Dasgupta et al. 2003) in antigen-presenting cells has been associated with immunosuppression. However, apparently paradoxical were the data observed with the blockade of B7-1 or B7-2 molecules that led to restoration of T-cell response and to increased IFN-γ and IL-4 production and parasite Clearance, respectively, in *L. chagasi* and *L. donovani*-infected mice (Waitumbi and Murphy 1993). The use of different ligand for B7 molecules, searched in sequence, explained this contradiction since there are two receptors for the B7 molecules, CD28 for T-cell activation and CTLA-4 for termination of T-cell activation. Indeed, blockade of CTLA-4 has led to the recovery of resistance against infection, suggesting expression of the CTLA-4 molecule during visceral leishmaniasis (Gomes et al. 1995; Waitumbi and Murphy 1993). Furthermore, it has been shown that the effect of CTLA-4 linkage resulted in the production of TGF-β, a factor that favors parasite growth within macrophages (Gomes et al. 1995). All of these data demonstrate a role of CTLA-4 in immunosuppression, favoring parasite growth, but there are other reports showing its role in the development of a Th1 response in *Leishmania major* infection in mice transfected with the CTLA-4 gene (Leifso et al. 2007). These paradoxical findings were elucidated in a review showing a dual role of the CTLA-4 molecule with activation of Th1 cells when T cells involved were naive, but with activation of Th2 cells when memory cells were involved (Gomes et al. 1995). This dual role of the CTLA-4 molecule is a crucial point to be further analyzed in an eventual study aiming at vaccine development. We should also emphasize the importance of TGF-β in susceptibility and immunosuppression since recent data have indicated it as one of the most important factors, maybe a determinant factor, leading to Th2 development through inhibition of T-bet in leishmaniasis.

Apoptosis of T cells have been reported in experimental visceral leishmaniasis. More than 40 % of CD4+ T cells from susceptible but not from resistant mice undergo apoptosis, accompanied by a significant decrease in IL-2 and IFN-γ secretion, and unaltered IL-4 secretion during *L. donovani* infection, findings that were also related to immunosuppression (Dasgupta et al. 2003). In addition, apoptosis was detected in inflammatory cells in the liver and spleen during *L. donovani* infection, but when the role of CD95-CD95 L was assessed using CD95 ligand-deficient mice, increased parasite growth, but no effect on apoptosis, was observed in the CD95 ligand deficient mice, suggesting a role of CD95 L in the control of parasite growth that is independent of host cell apoptosis (Alexander and Bryson 2005). Since apoptosis of host lymphocytes may have a role in immunosuppression leading to parasite growth, we addressed this question in the hamster model. It was observed that apoptosis of inflammatory cells in the liver and spleen of *L. chagasi*-infected hamsters that was induced by *Leishmania* antigen stimulation in the early period of infection. Based on these observations, we can speculate on the occurrence of selection of the lymphocyte populations, with apoptosis of *Leishmania* antigen-specific lymphocytes in the initial phase and survival of nonspecific ones in the later phase. This may explain the absence of a *Leishmania* antigen-specific lymphopro-liferative response throughout the study period but the presence of apoptosis only in the initial phase. These are some of the scattered lines of evidence about the role of apoptosis in immunosuppression that are still indirect and incomplete and that should be explored in the future.

Macrophages are not only the habitat of *Leishmania* but also the main cells involved in the leishmanicidal process. Survival of *Leishmania* depends on the integrity and supply or proliferation of macrophages in the lesion, besides a suppression of the leishmanicidal machinery. It was shown that *in vitro* infection of macrophages by *Leishmania* renders them resistant to apoptosis. We studied this phenomenon *in vivo* in hamsters with visceral leishmaniasis and observed that apoptosis is induced in macrophages by *L. chagasi* infection in the initial phase. However, as the infection progresses, apoptosis of macrophages disappears from both the liver and spleen, suggesting protection of macrophages by *Leishmania* infection. Soluble factors other than immunoglobulin have been involved in immunosuppression. Inhibition of concanavalin A-induced lymphoproliferative responses of splenic cells was seen in the presence of serum obtained during the acute phase of visceral leishmaniasis infection. Increased serum triglyceride levels were detected 60 days after infection and suppression was abolished when serum was delipidated.

2.1.5 Immunoprophylaxis

In the view of existing knowledge, lymphocytes, macrophages and antibody co-operation, the major components of the immune system, are necessary for protection against the leishmanial infection. The death of the parasite occurs either by stimulation of the sensitised lymphocytes through antigen load or by activating macrophages. The immunity observed after the therapeutic cure of kala-azar, suggests that vaccination against leishmaniasis is feasible and within the reach of conventional immunization methods. The advantage with leishmaniasis is that

- The infection produces a state of premunition (concurrent infection and immunity) and this would explain the long lasting protection to reinfection often seen in recovered individuals.
- It is possible to induce resistance to reinfection in animals by vaccination.
- The parasite promastigote can be grown in axenic culture easily and are therefore amenable to manipulations required for vaccine development.

There are two approaches towards development of *Leishmania* vaccine – Pragmatic and Systemic. The pragmatic approach involves trial of crude leishmanial components in animals and humans with or without BCG. The systemic approach requires identification, production and purification of protective immunogens, usage of adjuvant, carriers, mode of presentation and determination of protective immune responses. Both the approaches have made considerable progress towards developing antileishmanial vaccines. Several vaccine strategies have been developed (Modabber 1995). The "first generation" vaccines composed of killed parasites with or without adjuvant, are undergoing various stages of phase I (safety), II (reactivity) and Ill (efficacy) trials in humans conducted in Brazil, Venezuela, Iran. Vaccines with BCG as adjuvant have better protection as effective as chemotherapy. Second generation vaccine candidates, live vaccines, defined subunits and crude fractions are in preclinical development stages. Finally, protection results obtained with third-generation vaccines composed of cDNA encoding leishmanial antigens cloned into a eukaryotic expression vector are still in preliminary stage (Handman 1997).

2.2   Host Immune Response to *Leishmania* Infection

The initial immune response to parasite infection is elicited by host innate immunity that plays a major role in local inflammatory response and regulation of parasite dissemination. Rapid dissemination of amastigote is characteristic of infection in susceptible host and depends on infectious dose and species of the parasite. In C57BL/6 mice, local NK cells rapidly produce interferon-γ (IFN-γ) after activation by both parasite antigens and IL-12 leading to containment (Solbach and Laskay 2000). A protective anti-leishmanial adaptive immune response requires the presentation of appropriate antigens by antigen presenting cells, the induction and expansion of Th 1 lymphocytes and the activation of macrophages. Several data suggest an interference with MHC-II presentation of antigens of both parasitic and unrelated origins (such as OVA or β-galactosidase) on macrophages infected with *L.major*, *L.amazonensis*. Both *in vivo* and *in vitro* studies showed a significant difference in antigen presentation capacity of macrophages infected with promastigotes or amastigotes, the later being more inhibitory to parasite antigen presentation including "leishmania homolog of receptors for activated C kinase" (LACK). A specific interaction between parasites and MHC-II molecules in the PV including internalization of MHC-II molecules by amastigotes has been reported (Saha et al. 2006).

There is a decrease in B7-1 (Cd80) expression on macrophages in susceptible mice after infection resulting in decreased protective T cell response (Saha et al. 2006). Prolonged treatment with mAb to CD86 B7-2 decreased parasite burden and Th2 cytokines. Anti-CTLA-4 treatment in BALB/c mice suggested a major role of CTLA-4 in the early T cell unresponsiveness characteristic for chronic visceral leishmaniasis (Gomes et al. 1995; Saha et al. 2006). Induction of CD4+Th1/Th2 lymphocytes has been well documented in *L.major* infected mice showing elevated IL-4 mRNA but no IFN-γ in draining lymph node cells of non-healing BALB/C mice; while resistant C57BL/6 mice expressed IFN-γ, but only transiently for IL-4 (Heinzel et al. 1991). Further studies with parasite specific T cell lines that were either able or not to transfer resistance (Scott et al. 1987) also elucidated differences in cytokine production. Most studies suggest a critical role of initial cytokine milieu and corresponding Th cell subset development, B-cells and antibodies being of minor importance (Handman 1997; Scott 2003).

The responsiveness of Th1 cells to IL-12 in early phase of infection seems to be decisive for Th1 type development. Various studies also revealed role of TGF-β for suppressing IFN-γ driven processes and subsequent Th2 development including IL-12 unresponsiveness. The role of antigen-induced production of IL-10 is elicited due to its antagonistic effects on IFN-γ and association with susceptibility to visceralization of *L.major* in conjunction with 1L-4 (Heinzel et al. 1991). One of the few immuno-suppressive proteins, papLe22 from *L.infantum* was found to produce IL-l and aggravate the disease. IL-12 expression is influenced by Leishmania parasites in a stage dependent and species independent manner. Peripheral blood mononuclear cells from Kala Azar patients did not secret IL-12 after stimulated with leishmanial antigens *in vitro* (Ghalib et al. 1993). The production of superoxide (O^{2-}) and nitric oxide (NO) is the two major anti leishmanial effector mechanisms. Parasite-infected macrophages have a decreased capacity to produce oxygen radicals after stimulation with phorbol esters (Passwell et al. 1994). Several proteins like LPG, gp63 and TryP (Tryparedoxin peroxidase) have been shown to be inhibitory of O^2 production. Parasite destruction in macrophages is carried out most efficiently by NO, after being activated by T cell derived cytokines

Life long persistence of parasites in mammalian host cells is well established. Parasites could be detected by both PCR and culture in the draining lymph nodes, spleen and bone marrow (Aebischcr et al. 1993). The hiding parasites are fully virulent and induced fatal disease in 13AL13/c mice. Persistence was also found after clinical cure of human infections. Prevention of parasite replication above a certain threshold and thereby preventing exacerbation of the disease is an active process and performed by Th cell mediated activation of NOS2. The parasite usually persists in macrophage and dendritic cell but not in granulocytes or cells. The phenomena of post-kala-azar dermal leishmaniasis in form of macules and papules even after 20 years of successful treatment are another evidence of life long persistence in host cells. The phenomenon of visceralization has recently being found to be due to an amastigote specific protein A2 from visceralizing parasites (Zhang et al. 2006). Targeted gene disruption of A2 gene abolished visceralization. Immune response in different host after *Leishmania* infection is described below:

2.2.1 Human

VL is a potentially fatal disease and develops life long immunity against reinfection. Human VL caused by *L. donovani* or *L. infantum* is a severe disease with generalized spread of the parasites to reticuloendothelial system, such as spleen, liver and bone marrow. In the communities exposed to infection, individuals develop a strong cellular immunity in age-related fashion. There are several studies indicating that Th2 type response predominates during acute disease, such as suppression of T-cell reactivity to *Leishmania* antigens, predominance of endogenous IL-4 over IFN-γ production, and polyclonal B-cell activation resulting in hypergammaglobulinemia. In contrast with the *L. major* mouse model, however, other studies indicate that both Th1 and Th2 like cells appear to be activated during the course of the disease, as revealed by the simultaneous high levels of IFN-γ and IL-10 detected in patients. Furthermore, up-regulation of IL-10, rather than IL-4, appears to be a constant

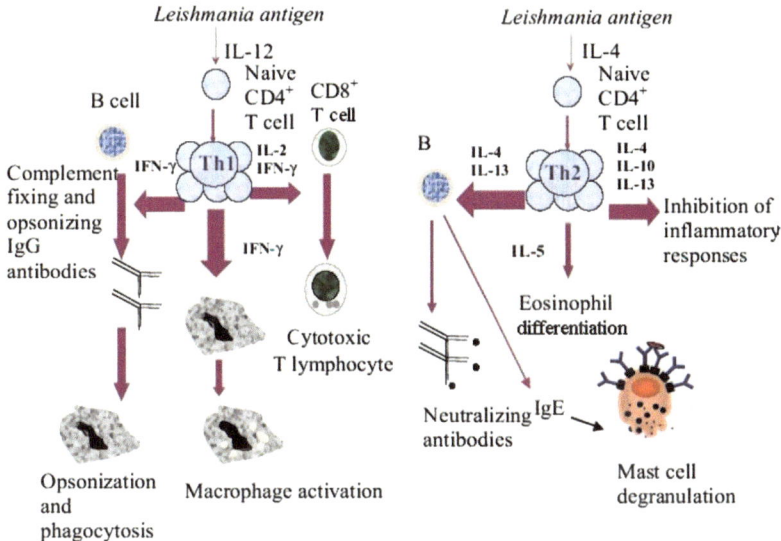

Fig. 2.2   Effector functions of Th1 and Th2 subsets of CD4+ helper T lymphocytes

feature in clinical disease. This cytokine has been associated to the immune suppression commonly seen in VL patients (Ghalib et al. 1993) and to the development of post-kala-azar dermal leishmaniasis, a skin manifestation caused by *L. donovani* after apparent VL healing. Taken together, above findings provide an evidence for the existence of a Th1/Th2 dichotomy in the T-cell response also in human VL. Interestingly, the outcome of the infection appears to be determined and regulated by the balance between the two parasite-specific T-cell populations (Kemp et al. 1993). Thus, in humans it is difficult to demarcate the responses leading either to visceral disease or to protective immunity with *L. donovani*. However, active VL also finds correlation with enhanced induction of IFN-γ, IL-2, IL-10 and IL-4. After cure, high levels of IFN-γ, IL-4 and IL-10 persist, suggesting a coexistence of Th1 and Th2 in kala-azar patients as well as in cured individuals (Kemp et al. 1993).

Which mechanism underlies the poor T-cell reactivity to *Leishmania* antigens? Much of this effect might be due to unbalanced production of IL-10. Indeed, elevated blood levels of IL-10 as well as high production of this cytokine (or its mRNA) by lymph node cells or PBMC from kala-azar patients appears to be a hallmark for this form of infection (Ghalib et al. 1993) (Fig. 2.2). In contrast to this cell populations collected from cured individuals fail to express this immunomodulatory molecule. IL-10 therefore, appears to constitute a major regulatory cytokine whose production may critically control the outcome of infection. Failure to produce IL-12 has similarly been associated with the active form of the disease (Ghalib et al. 1993). It is also suggested that unbalanced IL-10 production might play a role in progression of the disease towards the cutaneous form of the infection known as PKDL (Ghalib et al. 1993), which is observed in a high proportion of patients. Interestingly, PKDL appears to develop as soon as *Leishmania*-reactive T cells start to become detectable in peripheral blood following cure of the visceral infection, and it is also observed in the patients whose cells respond to *Leishmania* antigen *in vitro* by high IFN-γ release. Furthermore, histological examination of the lesions suggest that the development of skin lesions characteristic of PKDL involves an inflammatory component (Kemp et al. 1999).

In communities exposed to infection, individuals develop a strong cellular immunity in age-related fashion, as shown by cross-sectional studies on reactivity to intradermal leishmanial antigens (leishmanin skin test, LST). Sub clinical or asymptomatic cases identified by low-titer anti-*Leishmania* antibodies and surveyed prospectively, either may show anergy to intradermal antigens and develop disease thereafter, or convert to seronegative and become LST positive, suggesting spontaneous healing. During acute VL patients are unresponsive to LST, whereas after 6–12 months of successful treatment, the test converts to positive. Both spontaneous and drug-induced healing is thought to be followed by strong protective immunity. There are several studies indicating that Th2 type response predominates during acute disease, such as suppression of T-cell reactivity to *Leishmania* antigens, predominance of endogenous IL-4 over IFN-γ production, and polyclonal B-cell activation resulting in hypergammaglobulinemia (reviewed by (Rezai et al. 1978). Taken together, the above findings provide evidence for the existence of Th1/Th2 dichotomy in the T-cell response also in human leishmaniasis.

In humans, measurement of cytokines in culture supernatants of *Leishmania* antigen-activated PBMCs and T-cell clones have helped in determining the type of Th1 and Th2 immune responses being stimulated. Studies of tissue cytokine mRNA expression reveal a role for IL-10 in down regulating CD4$^+$ T cell responses and the involvement of IL-10 in disease pathology of *L. donovani* infections. However, active VL also finds correlation with the enhanced induction of IFN-γ, IL-2, IL-10 and IL-4. After cure, significant levels of IFN-γ, IL-4 and IL-10 persist, suggesting a coexistence of Th1 and Th2 in kala-azar patients as well as in cured individuals (Kemp et al. 1993; Sundar et al. 1998). The dichotomous Th1/Th2 cellular response seen in human VL is probably regulated by the balance of the parasite specific cellular populations (Kemp et al. 1993), which may play an important regulatory role in determining the outcome of the infection in humans. Thus, even in humans it is difficult to demarcate the responses that whether they are leading either to visceral disease or to protective immunity with *L. donovani*. Individuals with overt VL display a negative skin test response to *Leishmania* antigens. Furthermore, PBMCs from such individuals fail to proliferate or to produce IFN-γ when exposed to specific antigen *in vitro* (Passos et al. 2005). Interestingly, failure of PBMCs from recently infected infants to produce this cytokine is predictive of evolution of the disease towards classical, patent VL, whereas conversely, production of IFN-γ *in vitro* is indicative of maintenance of the infection at a sub clinical level. These observations point to the critical importance of early T cell activation patterns in determining the eventual outcome of infection.

2.2.2   *Mouse*

Till date, two host systems have been classified for studying *Leishmania* infection on the basis of susceptibility and resistance of the host. Murine models for experimental leishmaniasis are well established. Parasites are injected underneath the skin of the footpad. Most of the mice strains like C57BL/6, CBA/J, C3H or BIOD2 resist the infection with clinical cure within few months (Handman 1997); while BALB/c and all T-cell immunodeficient strains manifest a systemic VL leading to death (Howard et al. 1984). Resistance and susceptibility are closely related with the development of T-cell responses of Th1 or Th2 type, respectively.

It is well documented that Th1 immune response is the key event to prevent *Leishmania* infection. C57BL/6 mice mount early Th1 immune response and prevent the further growth of the parasite causes self-healing phenotype (Lehmann et al. 2000) whereas susceptible BALB/c strain mounts early Th2 response and results in non healing lesion and exaggeration of disease. Respective resistance and susceptibility of C57BL/6 and BALB/c strains depend not only on the Th1 and Th2 type of immune response of CD4+ T cells but also on the genetic background of the host. In general, the immune responses following infection of inbred mouse strains with viscerotropic *Leishmania* species, such as *L. donovani* or *L. infantum*, are similar to those observed in the *L. major* mouse model. However, BALB/c mice do not appear to exhibit a similarly high susceptibility to these parasites, since intravenous injection of visceral *Leishmania* results in a self-healing chronic infection. Furthermore, cytokine phenotypes elicited by viscerotropic *Leishmania* in this mouse model are not typical of a Th2-type response.

One of the most important advances in the last decade is our understanding of the Th1/Th2 paradigm, which has been best studied in the mouse model of *L. major*. While control of infection in resistant mice such as C3H and C57BL/6 is associated with the expansion of a Th1 subset of CD4 lymphocyte that secretes macrophage activating cytokines such as IFN-γ and IL-2 (Heinzel et al. 1991; Reiner and Locksley 1995), susceptibility is instead linked to expansion of the CD4 lymphocytes belonging to the Th2 subset that secrete IL-4, IL-5, IL-l0 and IL-13 (Heinzel et al. 1991). However, such a polarized Th1 or Th2 response has not been observed in murine models of *L. donovani* infections. Studies of C57BL/10 and B10-D2/N mice infected with *L. donovani*, the curing and non-curing models of VL, respectively, have shown that the Th2 type cytokine, IL-4, is not required for the susceptible phenotype (Kaye et al. 2004). However, Thl and Th2 cell associated cytokines have been reported in BALB/c mice infected with *L. donovani*. BALB/c mice infected with parasites of *L. donovani* develop either an initial susceptibility with spontaneous acquisition of resistance and control over parasite replication, or a progressive visceral disease mimicking infection of *L. donovani* in human. While the control of infection in the former model is known to be dependent, in part, on Thl cell responses characteristics of the T cell responses in the later have not been identified and provide a challenge for the development of a vaccine. To achieve a detectable visceral disease, often mice have to be challenged with a very large number of parasites, which is far from those inoculated during a natural infection. Recently, an intradermal murine model of VL has been explored for the establishment of a chronic infection pattern in susceptible BALB/c mice resembling a course of disease similar to that of human VL.

2.2.3   Hamster

The golden hamster is an experimental model for progressive visceral disease characterized by symptoms such as hyperglobulinemia, anemia and cachexia leading to the death of the animals. Interestingly, an inoculum (10^3 to 10^5) of *L. infantum* promastigotes could result in the development of both symptomatic and asymptomatic states with a spontaneous protective response (Soto et al. 1999). The fact that even a small inoculum which is close to the numbers seen during natural infection, could lead to disease, along with the unpredictable nature of immune state suggests that the various clinical and pathogenic features exhibited in hamsters are very much similar to those existing in human VL making it an excellent model of VL. Recent studies have shown that the mice model of *L. donovani* does not reproduce the features of active human VL like chronic fever, hepato-splenomegaly, pancytopenia and profound cachexia and have an ineffective anti-leishmanial cellular response (Melby et al. 2001). On the contrary, Syrian golden hamster model of active VL closely relate the human counter part as shown by relentless increase in visceral parasite burden, progressive cachexia, hepato-splenomegaly, pancytopenia, hypergammaglobulinemia and ultimately death. While the mice are either intrinsically resistant or susceptible to *Leishmania* infection and offer a well characterized genetic makeup, chiefly by the use of inbred, recombinant and naturally or experimentally mutated strains, hamsters provide an excellent model for an overtly susceptible host. Therefore, hamsters are used for histopathological studies, drug efficacy studies and vaccine studies despite the lack of fine immunochemicals that limit the mechanistic exploration of immune responses to *Leishmania* infection (Melby et al. 2001).

2.3   Cross-immunity Between Different *Leishmania* Species

One of the requirements of an 'ideal' anti-leishmanial vaccine is its effectivity against more than one *Leishmania* species in order to protect individuals in areas where CL and VL, for example, coexist. However, very little research has been done on the potential cross-protection induced by a vaccine derived from one *Leishmania* species against another. Initial studies using sequential infections with distinct species have suggested complex cross protection relationships (Santos et al. 2002). A few specific antigens have been tested for cross-protection with mixed success. For example, a DNA vaccine encoding the highly conserved LACK antigen cannot induce cross-species protection (Melby et al. 2001), whereas *L. donovani* promastigote antigen dp72 can induce protection against both *L. donovani* and *L. major* in mice (Rachamim and Jaffe 1993), A DNA vaccine encoding *L. amazonensis* P4 nuclease can also protect significantly against *L. amazonensis* and *L. major*, but different adjuvants (IL-12/HSP70-encoding plasmids) are required to obtain such protection (Medzihradszky et al. 2000). Although evidence of cross protection was observed in humans who were refractory to *L. mexicana* infection following recovery from *L. tropica*-caused leishmaniasis, such cross-protection is rare in humans. A high degree of variability in cross-immunity between the New World *Leishmania* species has been observed in humans as well as in experiments in simian models. Other studies showed that *L. major* DHFRTS knockouts were able to confer protection against *L. amazonensis*, and immunisation with heat-killed *L. donovani* promastigotes offered cross-protection against *L. major* challenge.

2.4 Adoptive Transfer of Immunity

During active infection with *L. donovani* in humans, CMI measured by DTH and LTT and the production of IFN-γ and IL-2 in response to leishmanial antigen is absent or reduced. After spontaneous or drug-induced cure, these parasite-specific responses recover indicating that healing of VL in humans is dependent primarily on the development of cell-mediated immunity. Furthermore, there are reports that protective immunity can be passively transferred to naive recipients with T-cell-enriched preparations from donor animals that have resolved infection (Rezai et al. 1978; Murray et al. 2000). The first description of a soluble leishmanial antigens stimulated T cell line transferring protective CMI against CL was done by Scott et al. (1987). Adoptive transfer of the splenocytes from BALB/c mice immunized with this preparation to syngenic recipient resulted in significant protection against *L. donovani* infection by transferring adoptive immunity with splenic T cells which were primarily responsible for mediating resistance to reinfection with Ethiopian *L. donovani* in immune C57BL/10 and irradiated BMO.D2 mice. A central role for T lymphocytes in the healing of cutaneous lesions has been further established by experiments showing that athymic nu/nu mice exhibited a generalized disease after infection with *L. major* and that resistance to infection in these highly susceptible mice could be achieved by the transfer of normal syngenic T lymphocytes (Handman 1997).

Chapter 3
Invade and Survival Strategy of *Leishmania*

Leishmania parasite has a digenic life cycle with phlebotomine sand flies (*Phlebotomus spp.*, *Lutzomyia spp.* and *Psychodopygus spp.*) as secondary inverte-brate vector host. In the sand fly, parasites initially reside within the peritrophic membrane of the midgut after ingestion of an infected blood meal. After release from macrophages, differentiation of amastigote to promastigote occurs with con-current synthesis of a thick glycocalyx coat composed of a variety of glycophospa-tidylinositol (GPI) anchor compounds. A layer of low molecular weight glycoinositol phospholipids (GIPLs) with barrier function is found below the glycocalyx. A lipo-phosphoglycan (LPG), containing a repeated polymer of disaccharide phosphate units, is the most abundant glycoconjugate (McConville and Ralton 1997). After 2 days, the peritrophic membrane is ruptured and the promastigotes attach to the mid gut wall through specific binding of LPG and rapidly divide. The division ceases after 4–7 days and parasite undergoes metacyclogenesis into infective metacyclic promastigote, which is unable of binding to the midgut wall due to structural altera-tion of LPG (Sacks et al. 1995). Then they migrate to the foregut and oesophagus, being suspended in saliva and ready to be inoculated. The insect cardiac valve is enzymatically damaged during the process that normally prevents reflux from gut to the pharynx (Schlein et al. 1992). The insect saliva promotes survival and develop-ment of promastigotes (Ghosh et al. 1995).

The sand fly rips up the epidermis while breaching the skin of the mammalian host and gains access to dermal capillaries. Small numbers (less than 100) of para-sites are regurgitated into the wound. However, experimental models need larger of number parasites for successful infection and low number of parasites (100–500) causes stable and protective immunity. The sandfly saliva itself has important dis-ease promoting effects, which is dependent on IL-4. The saliva promotes inflamma-tion, inhibits oxidative metabolic processes and antigen presentation by macrophages *in vitro*. A protein phosphatase inhibitor and a vasodilator Maxadilan were identi-fied from salivary glands as the active principle. Most of the extracellular promasti-gotes, before entry into macrophages, are killed by complement factors. However, some survive due to resistance to complement-mediated lysis, which is achieved by

A. Kumar, *Leishmania and Leishmaniasis*, SpringerBriefs in Immunology 3, DOI 10.1007/978-1-4614-8869-9_3, © Springer Science+Business Media New York 2013

LPG-associated spontaneous shedding of C5b–C9 complexes from the parasite surface. The metacyclic promastigotes also possess a serine/threonine protein kinase
(LPK-1), which inactivates C3, C5 and C9 by phosphorylation. At the same time
Leishmania also expresses a zinc proteinase called gp63 (also termed leishmanolysin), which converts opsonic complement factor C3b to iC3b promoting uptake of
the parasites by host cells having iCR3 (CD11b/CDI8).

The surviving *Leishmania* adheres to the cells of monocyte/macrophage lineage
including dendritic cells and Langerhans cells (Moll et al. 1995) and also human
granulocytes (Dominguez and Torano 1999). The adherence is mediated by several
receptors like mannose–fucose receptor, the fibronectin receptor and the receptor
for C-reactive protein. Complement receptors type I (CR1, CD3S) and type III
(CR111, CD11b/CD18) bind to complement components attached to the plasma
membrane of the parasite, followed by internalization either by "coiling phagocytosis" (Rittig et al. 1998) or by "zipper like interactions. The phagosomes fuse with
endocytic organelles and form a parasitophorous vacuole (PV) that contains hydrolases, cathepsins and β-glucuronidase. In the PV, the promastigotes transform into
non-motile amastigotes within two days. Amastigotes do not synthesize LPG but
contain related compound proteophosphoglycan, acid phosphatase and GIPLs. The
numerous small PVs of *L. donovani* contain only one or few amastigotes in each.
The acidic pH is maintained by an H^+-ATPase of host cell origin and also by a P type
H^+-ATPase of parasite cell membrane.

The pathology of *Leishmania* infection is determined not only by the parasite species, but also by host genetics and immune factors. Most of the experimental immunological data come from mouse models and less is known about the immunology of
human leishmaniasis. Although mouse models have been used for the study of both
CL and VL, they more closely reflect the situation in human cutaneous leishmaniasis
than visceral disease. In the case of CL, effective protection against infection has
been largely attributed to the development of a potent $CD4^+$ Th1 type immune
response, characterized by the production of IL-12 and IFN-γ, which subsequently
mediates macrophage activation, nitric oxide production and parasite killing
(Alexander and Bryson 2005). A clear-cut polarization of T helper cell responses is
not evident in human leishmaniasis which shows a mixed Th1 and Th2 immune
response. The ability of the infected individual to mount a Th1 response is considered to be partially responsible for the observed differences in the clinical picture of
leishmaniasis. However, Th2 cell mediated responses have not been unequivocally
associated with the failure to mount a protective response, and therefore causing
long-lasting cutaneous or systemic infection. The disease phenotype may be attributed to the *Leishmania* species causing the disease. In recent years, the involvement
of $CD8^+$ T cells has also been shown to play an important role in immunity against CL.

Recent studies on the generation and maintenance of central memory (CM) and
effector memory (EM) $CD4^+$ T cells during cutaneous *Leishmania* infection shed
new light on the design of effective vaccination strategies against *Leishmania* (Scott
2005). In the murine model of disease, it has been suggested that the constant presence of live parasites is required for maintaining EM CD4+ T cells, but might not
be essential for the maintenance of CM $CD4^+$ T cells. However, the importance of
persistent infection for maintaining an effective long lasting protective response is

controversial. Since vaccines need to generate immunological memory, a better understanding of the formation and maintenance of CM and EM CD4$^+$ T cells in both animal models and human disease will be critical for their development.

3.1 Host–Parasite Interaction

Leishmaniasis is an excellent example of a complex parasite–host interaction. *Leishmania* promastigotes bind to some of the surface molecules like complement receptor 1 and 3 (CR1&3) and C3b of macrophage before they are internalized. CR1 constitutes the major macrophage ligand for mature promastigotes, though additional parasite surface glycoprotein (*e.g.*, gp63 membrane protease) and other macrophage receptors (*e.g.*, CR3, mannose fucose receptor) have been implicated in various studies. Once internalized, promastigotes transform into intracellular amastigotes. Amastigotes replicate by binary fission, eventually rupturing the macrophage and spreading to uninfected cells. The internalization pathway of amastigotes to the macrophages is poorly defined. The identification of natural antibodies coating amastigotes *in vivo* suggested that macrophage FcIg and CR3 receptor might contribute to phagocytosis. *Leishmania* promastigotes are covered with a dense surface glycocalyx, composed largely of molecules attached by glycosylphosphatidylinositol (GPI) anchor. These GPI anchored molecules include proteins such as the parasite surface protease gp63 and proteophosphogycans (PPGs). The most abundant constituent is a large GPI-anchored phosphoglycan called lipophosphoglycan (LPG). LPG and gp63 account for the virulence of the parasite. LPG has been implicated in many steps required for the establishment of macrophage infection and for the survival in insect vector. LPG does not play a role in the amastigotes stage; however, amastigotes continue to make structurally related glycoconjugates. On the other hand, gp63 also helps the parasite to enter in the host cells and for its survival. As an endoproteinase with a broad substrate spectrum, gp63 has the potential to degrade immunoglobulins, complement factors, and lysosomal proteins. Its proteolytic activity at *p*H 4 bears apparent relevance to the survival of amastigotes in the acidic environment of macrophage phagolysosomes.

Further molecular basis of this mechanism remains to be deciphered. Toll-like receptors (TLRs) were discovered originally in *Drosophila* but later on they have also been recognized on mammalian cells including macrophages and dendritic cells. TLRs are the components of the innate immunity and induce nonspecific immune response. Since TLRs are present on the surface of macrophages, the cells that play host to *Leishmania*, There is strong possibilities that *Leishmania* may also bind to TLRs and can switch on the early genes of the host to regulate the parasite and host interactions but the mechanism is not known. After internalization of organisms into phagosomes, secondary lysosomes are fused to form the complete parasitophorus vacuole. Metacyclic forms rapidly transform into intracellular amastigotes. This transformation takes the shedding of the promastigote LPG that migrated to the surface of the infected macrophages. LPG inhibited the respiratory burst, a natural process that occurs after phagocytosis, and the hydrolytic activity of the lysosomal enzymes, possibly through chelation of calcium and inhibition of protein kinase C. Various markers have been used to study the intracellular compartment in which amastigotes replicate in macrophages. For example, a proton ATPase and LAMP-1. With time, the vacuole matures to a late endosomal compartment. Association of MHC class II molecules to parasitophorus vacuole suggesting a mechanism by which the immune response to this intracellular organism becomes class II and CD4$^-$ dependent. Recognition of antigenic peptides of the parasite through the class II pathway was revealed by experiments using mice without MHC II or β2-microglobulin genes that are deficient in MHC class II/CD4 cells or MHC class I/CD8 cells, respectively. MHC class II deficient mice suffered fatal, uncontrolled infection whereas MHC class I-deficient mice controlled infection with *L. major* in a manner similar to normal littermates, suggesting that CD4$^+$ but not CD8$^+$ T cells are required for controlling *L. major* infection.

3.2   Antigen Presentation and Immune Components Involved

Neutrophils or polymorphonuclear neutrophils (PMNs), the first cells to migrate to the site of infection or injured tissue, function as a primary effector or phagocytic cells, phogocytosing *Leishmania*. Foreign particles are destroyed by proteolytic enzymes stored in the special granules and by production of reactive oxygen species. *Leishmania* phagocytosed neutrophils start secreting the chemokines like IL-8 essential to bring the more neutrophils at the site of infection. Two to three days later, the second wave of cells, monocytes/macrophages, enters the site of infection. After infection with *L. major*, the chemokines MIP-2 and KC (the functional murine homologues of IL-8) are very rapidly produced in the skin. Dendritic cells (DCs) are potent antigen presenting cells and can induce T cell activation efficiently. It has also been shown that DCs are the source of different cytokines such as IL-12, IL-10 and IFN-γ. Incubation of *Leishmania* promastigotes with dendritic cells induced early IL-12 production *in vitro*, which might be contributed from the pre existing pool of IL-12 p70 which was secreted soon after ligation of any microbial product, suggesting the role of DCs in the initiation of T cell immune response in *Leishmania* infection. It is also reported that uptake of *Leishmania* amastigotes by skin derived DCs induces IL-12 p70, upregulates of costimulatory molecules and vaccinates against *L. major* infection; in marked contrast, *L. major* inhibits IL-12 production in macrophages. It was shown the CC chemokines receptor (CCR) 2-/- mice are defective in DCs migration from marginal zone of lymph node to T cells area and markedly impaired in antigen specific T cell activation and make resistant mouse strain susceptible for *L. major* suggesting the regulatory role of chemokines receptors in parasitic infections. Another study suggested that down regulation of CCR7 by *L. donovani* impaired the DCs migration and used by the parasite as an immune evasion strategy, contributing to disease progression. Dendritic cells are potent candidate for immunotherapy of leishmaniasis. Upon loading with microbial antigen (Ag) and adoptive transfer, DCs are able to induce immunity to infections.

3.3 Survival and Evasion Strategies of Leishmania in Macrophages

Most of the extracellular promastigotes, before entry into macrophages, are killed by complement factors. However, some survive due to resistance to complement-mediated lysis, which is achieved by LPG-associated spontaneous shedding of C5b–C9 complexes from the parasite surface. The metacyclic Leishmania also possess a serine/threonin protein kinase (LPK-1), which inactivates C3, CS and C9 by phosphorylation. At the same time Leishmania also expresses a mctzincin zinc proteinase called gp63 (also termed leishmanolysin), which converts opsonic complement factor C3b to iC3b promoting uptake of the parasites, by host cells having iCR3 (CD11b/CDI8) (Sacks et al. 1987). The surviving Leishmania adheres to cells of the monocyte/macrophage lineage including dendritic cells and Langerhans cells and also human granulocytes. The adherence is mediated by several receptors like mannose–fucose receptor, the fibronectin receptor and the receptor for C-reactive protein. Complement receptors type I (CR1, CD3S) and type III (CR111, CD11b/CD18) bind to complement components attached to the plasma membrane of the parasite, followed by internalization either by "coiling phagocytosis" or by "zipper like interactions. The phagosomes fuse with endocytic organdies and form a parasitophorous vacuole (PV) that contains hydrolases, cathepsins and β-glucuronidase. In the PV, the promastigotes transform into non-motile amastigotes within 2-days. Amastigotes do not synthesize LPG but contain related compound proteophosphoglycan, acid phosphatase and GIPLs. The numerous small PVs of *L. donovani* contain only one or few amastigotes in each. The acidic pH is maintained by an H^+ATPase of host cell origin and also by a P type H^+ATPase of parasite cell membrane (Saha et al. 2006).

Macrophage activation by T-cell-derived cytokines is required to establish control of cellular infection and progressive disease. In these cells, *Leishmania* growth inhibition or killing is mediated by production of nitric oxide from inducible nitric oxide synthase (iNOS). Transcription of this enzyme is induced and enhanced by interferon IFN-γ a key cytokine in the *Leishmania* control, and by other synergic cytokines such as tumor necrosis factor TNF-α or interleukin (IL)-2. Macrophage activation results from a network of up- or down-regulating chemokine signals which modulate immune cell proliferation, activation, and expansion. The initial immune response to parasite infection is elicited by host innate immunity that plays a major role in local inflammatory response and regulation of parasite dissemination. Rapid dissemination of amastigote is characteristic of infection in susceptible host that depends on infectious dose and species of the parasite. In C57BL/6 mice, local NK cells rapidly produce IFN-γ after activation by both parasite antigens (Stobie et al. 2000).

A protective anti-leishmanial adaptive immune response requires the presentation of appropriate antigens by antigen presenting cells, the induction and expansion of Th 1 lymphocytes and activation of macrophages. Several data suggest an interference with MHC-II presentation of antigens of both parasitic and unrelated origins (such as OVA or β-galactosidase) on macrophages infected with *L. major*, *L. amazonensis*. Both *in vivo* and *in vitro* studies showed a significant difference in antigen presentation capacity of macrophages infected with promastigotes or amastigotes, the later being more inhibitory to parasite antigen presentation (Kima et al. 1996) including "leishmania homolog of receptors for activated C kinase" (LACK). A specific interaction between parasites and MHC-II molecules in PV including internalization of MHC-II molecules by amastigotes has been reported (Kemp et al. 1993).

There is a decrease in B7-1 (Cd80) expression on macrophages in susceptible mice after infection resulting in decreased protective T cell response. Prolonged treatment with mAb to CD86 B7-2 decrease parasite burden and Th2 cytokines. Anti-CTLA-4 treatment in BALB/c mice suggested a major role of CTLA-4 in the early T cell unresponsiveness characteristic for chronic visceral leishmaniasis. Induction of CD4+Thl/Th2 lymphocytes has been well documented in *L.major* infected mice showing elevated IL-4 mRNA but no IFN-γ in draining lymph node cells of non-healing BALB/C mice; while resistant C57BL/6 mice expressed IFN-γ, but only transiently for IL-4. Further studies with parasite specific T cell lines that were either able or not to transfer resistance also elucidated differences in cytokine production. Most studies suggest a critical role of initial cytokine milieu and corresponding Th cell subset development, B-cells and antibodies being of minor importance (Saha et al. 2006; Scott et al. 1988).

The responsiveness of Th1 cells to IL-12 in early phase of infection seems to be decisive for Th1 type development. Various studies also revealed role of TGF-β for suppressing IFN-γ driven processes and subsequent Th2 development including IL-12 unresponsiveness. The role of antigen-induced production of IL-10 is elicited due to its antagonistic effects on IFN-γ and association with susceptibility to visceralization of *L.major* in conjunction with 1L-4. One of the few immuno-suppressive proteins, papLe22 from *L. infantum* was found to produce IL-1 and aggravate the disease. IL-12 expression is influenced by Leishmania parasites in a stage dependent and species independent manner. Peripheral blood mononuclear cells from Kala Azar patients did not secret IL-12 after stimulated with leishmanial antigens *in vitro*. Amastigotes also inhibited IL-12 secretion along with CD40 cross-linking and cognate interaction with Th1 cells (Weinheber et al. 1998). IL-12 is indispensable for protective immunity probably through NK cell activity, IFN-γ production and IL-4 suppression by IFN-γ independent way (Ghalib et al. 1993).

The production of superoxide (O^{2-}) and nitric oxide (NO) is the two major anti leishmanial effector mechanisms. Parasite-infected macrophages have a decreased capacity to produce oxygen radicals after stimulation with phorbol esters. Several proteins like LPG, gp63) and TryP (Tryparedoxin peroxidase) have been shown to be inhibitory of O^2 production. Parasite destruction in macrophages is carried out most efficiently by NO, after being activated by T cell derived cytokines. Most data suggest that Leishmania inhibits NOS2 activity in the early stages of infection, while in later stages it increased NOS2 production. Macrophages infected with *L.major in vitro* upregulated FAS in response to IFN-γ and became susceptible to Th1 cell induced apoptotic death limiting the number of host cells (Leifso et al. 2007).

Lifelong persistence of parasites in mammalian host cells is well established. Parasites could be detected by both PCR and culture in the draining lymph nodes, spleen and bone marrow. The hiding parasites are fully virulent and induced fatal disease in 13AL13/c mice. Persistence was also found after clinical cure of human infections. Prevention of parasite replication above a certain threshold and thereby preventing exacerbation of the disease is an active process and performed by Th cell mediated activation of NOS2. The parasite usually persists in macrophage and dendritic cell but not in granulocytes or cells. The phenomena of post-kala-azar dermal leishmaniasis in form of macules and papules even after 20 years of successful treatment are another evidence of lifelong persistence in host cells. The phenomenon of visceralization has recently being found to be due to an amastigote specific protein A2 from visceralizing *Leishmania* parasites (Zhang et al. 2001).

Chapter 4
Drugs and Diagnosis for Leishmaniasis

4.1 Drugs Against Leishmaniasis

4.1.1 Pentavalent Antimonials [Sb(V)]

Antimonials were first introduced in 1945 and these remain the effective drugs for all forms of leishmaniasis for more than 60 years. These are available as branded products, meglumine antimoniate (Glucantime) and sodium stiboguconate (Pentostam) (Fig. 4.1) and in the generic form, Sodium Antimony Gluconate (Albert David Ltd., Kolkata). The drugs are given by i.v. and i.m. routes and their efficacies are equivalent at similar doses (Sundar and Chatterjee 2006). The major concern for antimonials is that they exert toxic effects like arthralgia, nausea, abdominal pain, pancreatitis and cardiotoxicity. Moreover, treatment require hospitalization for a long (3–4 weeks) period. In Bihar state of India, over 60 % of previously untreated patients are unresponsive to Sb (V) rendering the drug useless for routine use (Sundar et al. 2006, 2007). To date, the precise mechanism of action of antimonial drugs remains an enigma. A general consensus is that Sb(V) acts upon several targets that include influencing the bioenergetics of Leishmania parasites by inhibiting parasite glycolysis (Mottram and Coombs 1998; Berman and Gallalee 1987), fatty acid β-oxidation (Berman et al. 1989) and inhibition of ADP phosphorylation (Berman and Gallalee 1987). It has also been reported to cause non specific blocking of SH groups of amastigote proteins and cause inhibition of DNA topoisomerase I. Recent evidences have shown that both Sb(III) and Sb(V) mediate DNA fragmentation in Leishmania species, suggesting that antimony kills the parasite by a process reminiscent of apoptosis.

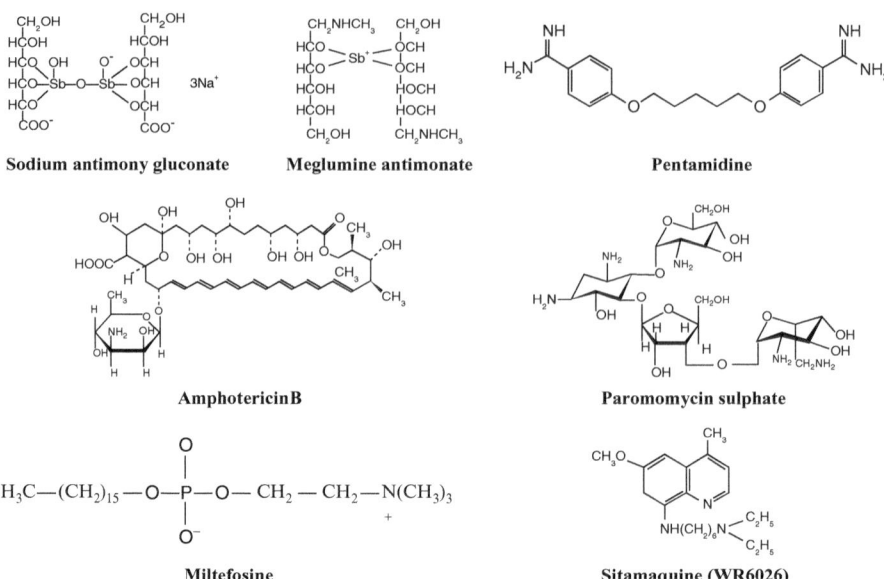

Fig. 4.1 Antileishmanial drugs currently available for treatment/under various stages of clinical trials

4.1.2   Amphotericin B

Amphotericin B is an antifungal macrolide antibiotic isolated from Streptomyces nodosus in 1956. Its antileishmanial activity was first shown in the early 1960s attributed to its selective affinity for ergosterol vis-a-vis cholesterol (Ramos et al. 1996). Amphotericin B, at a dose of 0.75–1.0 mg/kg for 15–20 infusions either daily or on alternate days, has consistently produced cure rates of about 97 % and is now the drug of choice in north Bihar (Sundar et al. 2002). It selectively inhibits the membrane synthesis of the parasite and causes holes in the membrane, leading to parasite death (Ramos et al. 1996). At present, three lipid formulations of Amphotericin B are available: liposomal Amphotericin B (AmBisome), Amphotericin B lipid complex (ABLC; Abelcet®) and Amphotericin B cholesterol dispersion (ABCD; Amphotec™). These preparations have been tested successfully for VL in India, Kenya and Brazil (Thakur et al. 1996; Sundar 2001). The lipid formulations had an upper edge as they produced distinctly lower toxicities, notably the absence of nephrotoxicity and significantly lower infusion reactions. Major limiting factors include an almost universal occurrence of infusion based reactions like high fever with rigor and chills, thrombophlebitis and occasional serious toxicities like myocarditis, severe hypokalaemia, renal dysfunction and even death.

4.1.2.1 Formulations of Amphotericin B

- Fungizone®: Amphotericin B deoxycholate. Contains no lipids.
- Emulsification of Fungizone® in Intralipid 20 %: little reduction of toxicity
- AmBisome®: L-AmB: incorporation in liposomes (vesicles).
- Abelcet®: ABLC or Amphotericin B Lipid Complex. Microscopically small ribbon-like membranes formed by complexing with phospholipids.
- Amphotec®: ABCD (= Amphocil®) Amphotericin B Colloidal Dispersion: AmB-cholesteryl sulphate forms disc-shaped structures.

4.1.3 Pentamidine

It is an aromatic diamidine and proven to be useful in Sb(V) resistant VL cases in India in the late 1970s and early 1980s and a cure rate of 98.8 % was reported without any relapse (Jha 1983). Its leishmanicidal activity is possibly mediated via its influence on polyamine biosynthesis and the mitochondrial membrane potential. Pentamidine has been abandoned as second-line treatment for VL because of toxicity (myalgia, nausea, headache, hypoglycaemia, irreversible insulin dependent diabetes mellitus and death) and declining efficacy.

4.1.4 Miltefosine (Hexadecylphosphocholine)

Miltefosine, initially developed as an anticancer drug, is the first effective oral treatment of VL. Its antileishmanial activity was initially discovered in the mid-1980s and since then its efficacy has been demonstrated in several in vitro and in vivo experimental models (Croft et al. 2006; Croft and Engel 2006). These findings led to clinical trials and registration of miltefosine in India in March 2002 for oral treatment of VL (Sundar et al. 2003). Its mode of action has been established to be apoptosis like cell death. Its adverse effects were mild to moderate gastrointestinal disturbances that include vomiting (40 %) and diarrhea (15–20 %) of patients. As Miltefosine is teratogenic, it is contraindicated in pregnancy. A potential problem is prolonged half-life of miltefosine (150–200 h) that raises concerns for emergence of drug resistance.

4.1.5 Paromomycin (Aminosidine)

Paromomycin (aminosidine) although developed in the 1960s as an antileishmanial agent, it remained neglected until the 1980s when topical formulations were found to be effective in CL and a parenteral formulation for VL was also developed. One World Health, the Bill and Melinda Gates Foundation, Gland Pharma Limited,

IDA Solutions and WHO/TDR partnered to develop Paromomycin injection as a public health tool to be sold on a not-for-profit basis at a very low price. In phase III clinical trial, 94.6 % patients treated with its injection were cured of VL. Paromomycin injection was approved on August 31, 2006 for treatment of VL in India (Sundar et al. 2007). In *L. donovani,* paromomycin promoted ribosomal subunit association of both cytoplasmic and mitochondrial forms, following low Mg^{2+} concentration.

4.1.6   Sitamaquine (WR 6026)

Sitamaquine, an orally active 8-aminoquinoline analog, was originally developed as WR6026 by the Walter Reed Army Institute of Research in collaboration with GlaxoSmithKline. Its antileishmanial activity was first identified in 1970s (http://www.wrair.army.mil). Although animal studies showed very encouraging results, human trials done on Kenyan patients did not find it more than 50 % effective (Chapman et al. 1981; White et al. 1990). Sitamaquine is rapidly metabolized, forming desethyl and 4-CH_2OH derivatives, which might be responsible for its activity. Toxicity appears to be relatively mild as it causes mild methemglobinaemia.

4.1.7   Pamidronate

It is a bisphosphonate drug typically used in the treatment of osteoporosis, is effective against experimental cutaneous leishmaniasis. Several bisphosphonates have significant activity against *Leishmania donovani* in vitro, and several are potent inhibitors of bone resorption and in clinical use for the treatment of osteoporosis and Paget's disease. Action on bone is based on binding of bisphosphonate moiety to the bone mineral and inhibition of the osteoclast's enzyme FPPS (farnesyl pyro-phosphate synthase). Expressed FPPS is also potently inhibited by bisphosphonates in the trypanosomatid parasite *Trypanosoma cruzi*, in which FPPS is postulated to be the major target of bisphosphonates. It is possible that currently approved clinical regimens of the drug are not high enough to cure human cutaneous leishmaniasis. Pamidronate could be a useful lead compound in the synthesis of new drugs against this disease.

4.2   Cutaneous Leishmaniasis (CL)

4.2.1   Distribution

Approximately 90 % of all cases of cutaneous leishmaniasis now occur in Iran, Syria, Saudi Arabia, Afghanistan, Peru and Brazil.

4.2.2 Clinical Features

Various forms are clinically distinguished, the most important of which are:

- **Localised cutaneous leishmaniasis**: skin ulcers that heal very slowly or nodular lesions, limited in extent and number. These chronic sores have regional names: clou de Biskra in Algeria and Aleppo boil in Syria.
- **Diffuse cutaneous leishmaniasis**: cutaneous nodules and plaques that do not ulcerate but can sometimes spread over the entire body.
- **Recurrent cutaneous leishmaniasis** : Recurring cutaneous leishmaniasis seldom occurs (Iraq, Iran). This disease, also known as leishmaniasis recidivans leads to significant tissue damage. Parasites are very difficult to detect in these very chronic lesions.
- **Localised cutaneous Leishmaniasis:** Satellite lesions can occur. Spontaneous healing often occurs after 6–12 months, resulting in a depressed scar.

4.2.3 Diffuse Cutaneous Leishmaniasis

Diffuse cutaneous leishmaniasis (DCL) is a diffuse affection of the skin with extensive non-ulcerative nodules and is a very chronic disease. It is sometimes followed by chronic lymphoedema of an affected part of the body. This disease is poorly understood, but is probably caused by a diminished resistance to the parasite. This immunosuppression is possibly brought about by the parasite itself. In East Africa diffuse cutaneous leishmaniasis is often caused by *L. aethiopica* and in the New World frequently by *L. mexicana*. If there are generalised cutaneous lesions the condition has to be differentiated from lepromatous leprosy, keloids, neurofibromatosis and post kala azar dermal leishmaniasis (PKDL). Due to the low resistance of the patient very numerous amastigotes are present skin smears are always positive. Treatment is difficult, as the patient's immune system itself is functioning poorly. Differentiation from PKDL is important, as the latter can still be treated reasonably well. In Sudan 1 case of DCL is found for every 100 cases of localised cutaneous leishmaniasis. The incidence varies greatly from district to district. It occurs frequently in South America, but in contrast to this it does not occur in India.

4.2.4 Recurring Cutaneous leishmaniasis

Attempts should be made to detect the parasite microscopically in a biopsy or smear from the edge of the wound. The biopsy will, if possible, be divided up for pathology (seldom available, not very sensitive, is principally used more for exclusion of another cause) and cultures (bacteria, mycobacteria, fungi, *Leishmania*) and an impression preparation should also be made. Lesions on the face can be injected with 0.1 ml physiological saline and aspirated again while moving the small, thin needle back and forth in the skin. Serology is usually negative. Differential diagnosis includes ulcers due to mycobacteria, cutaneous diphtheria, tertiary syphilis, yaws, cutaneous carcinoma and deep or subcutaneous mycosis. Acid fast bacilli can be made visible using the method of Ziehl-Neelsen. Field sore (cutaneous diphtheria) and tropical ulcers (fusobacteria + *Borrelia*) are painful, particularly in the early phase.

4.2.5 Localised Cutaneous Leishmaniasis

After the bite by a sandfly infected with *L. tropica*, there is an incubation period of a few weeks or months, occasionally years. There is initially a small papula and usually only a single lesion, though sometimes there are several. This slowly spreads, can remain completely dry, become warty or nodular or develop into a painless, sharply delineated ulcer surrounded by purplish raised border. Recurring cutaneous lesions—possibly with severe disfigurements—occasionally occur. There is usually immunity to any subsequent infection with the same organism. In infection with *L. major* (mainly rural infections, particularly from a rodent reservoir) the lesions are usually larger and develop more quickly, hence the name. There is a greater tendency to local spreading via the lymphatics and have to be distinguished from sporotrichosis. The lesions will eventually spontaneously heal with scar formation. In South America the lesions often have their own local names and clinical expressions. Hence in Peru they are called "uta" (a solitary ulcer or a few restricted lesions brought about by *L. peruviana*, frequently on the face). In Guyana they are known as "bush yaws" or (French) "pian bois" (*L. guyanensis*) with rasberry-like lesions that resemble yaws. In Yucatan, Mexico an ulcer on ear (caused by *L. mexicana*) is know as ulcer.

4.2.6 Treatment

The response to treatment varies according to the species. Drugs for systemic and topical treatment can be used. There is an urgent need for better and cheaper drugs.

- Physical methods: cryotherapy (liquid nitrogen) for 15–20″, repeated 2–3 times. Blistering will occur.
- Application of local heat (e.g. infrared lamp). Heat-induced skin bullae are common.
- Ointment with 15 % paromomycin and 12 % methylbenzethonium chloride in soft white paraffin (e.g. Leishcutan® ointment). Urea can be added as a keratolytic. Twice daily application is advised, for duration of 20–30 days.
- Skin infiltration with pentavalent antimony with a fine gauge needle. Blanching of the lesions should be obtained. Treatment is repeated every 5–7 days, in general 2–5 times.

- Imiquimod crème (Aldara®). This immunomodulator activates macrophage killing of *Leishmania* amastigotes, but is best used in combination with systemic meglumine antimonate. Experience with this drug is limited. Local application of imiquimod crème (250 mg, 5 % weight/volume), i.e. one individual packet every other day × 20 days is possible.
- Pentavalent antimonials (meglumine antimonate [85 mg Sb/ml, IM] or sodium stibogluconate [100 mg/ml, IV]. Duration of treatment is not standarised (e.g. 14–28 days).
- Pentamidine. First line against *L. guyanensis* (French Guyana). Check glycaemia. Several treatment schemes exist and the cure rate is dose-dependent. Some short-courses use 1,200 mg as a total dose. In Guyana 3 mg/kg/day every other day is often used (4 injections).
- Imidazoles, triazoles. Fluconazole promising against *L. major*. Ketoconazole 600 mg per day × 28 days is moderately effective for *L. mexicana*, but much lower against *L. braziliensis*.
- Miltefosine. Still experimental.
- Amphotericine B and its liposomal formulation
- Allopurinol. Not as monotherapy, but associated with antimonials for *L. panamensis*.

Glucantime® (meglumine antimonate) or Pentostam® (stibogluconate) can be injected intralesionally (that is, into the edge of the lesion itself) as a treatment for cutaneous leishmaniasis. These can be given parenterally for extensive lesions. Varying results have also been reported with allopurinol (Zyloric®), which can be given orally. Topical treatments with heating (40–42 °C for 12 h), freezing with liquid nitrogen and paromomycin ointment (15 % aminosidine in methylbenzethonium BD × 15–30 days) have been used with varying success. Itraconazole (Sporanox®) gave good results in initial studies, but is still controversial. Ketoconazole is sometimes used, but is use is often complicated by hepatotoxicity, abdominal pain and nausea. Imiquimod 5 % cream (Aldara®) is an immunomodulating substance, initially used for warts caused by papilloma virus. Its use in cutaneous leishmaniasis is still experimental. The treatment of diffuse cutaneous leishmaniasis caused by *L. aethiopica* is problematical, as this parasite is less sensitive to Glucantime®. Pentamidine can be used against *L. aethiopica*. A dose of 4 mg/kg/week which has to be continued for at least 4 months after disappearance of the parasites from the skin is an acceptable guideline here. Parenteral aminosidine sulphate is another therapeutic possibility. This is an antibiotic that is obtained from *Streptomyces chrestomyceticus*. It is an aminoglycoside and is thus potentially nephro- and ototoxic. It is chemically identical to paromomycin, which is obtained from a related *Streptomyces* strain. The compound is not resorbed from the intestine. Recurrences are frequently seen with aminosidine given as monotherapy. Aminosidine is, however, synergistic with stibogluconate and a permanent remission can be obtained with the combination of aminosidine with Glucantime® or Pentostam®. The dose is 14 mg/kg/day IM to be continued for up to 60 days after all parasites have been eliminated. The total treatment period takes 6 months or more. Good results were obtained with amphotericin B.

4.3   Mucocutaneous Leishmaniasis (MCL)

4.3.1   Distribution

At present 90 % of all mucocutaneous leishmaniasis occurs in Bolivia, Peru and Brazil. Illustrations of skin lesions and disfigurements suggestive of leishmaniasis are encountered on pre-Inca earthenware. These indicate that the disease was already in existence in Peru and Ecuador in the first century AD. Texts dating from the fifteenth to sixteenth century Inca period and the Spanish conquest mention the risk of cutaneous ulcers in seasonal farmers. Espundia was also described as "white leprosy".

4.3.2   Clinical Features

When skin and mucosae are affected the disease is known as mucocutaneous leishmaniasis. This is very rare in East Africa but frequent in South America, where it is known as "espundia". After an initial skin lesion, that slowly but spontaneously heals, chronic ulcers appear after months or years on the skin, mouth and nose, with destruction of underlying tissue. Tissue destruction with disfigurement can be very severe. Parasites are usually rare in the lesions. A substantial part of the disfigurement is possibly due to immunological mechanisms. One hypothesis is a relationship between the occurrence of mucocutaneous lesions and the presence of certain alleles of polymorphic tumour necrosis factor α and β genes.

4.3.3   Diagnosis

The lesions often contain few parasites. Diagnosis is sometimes made solely on a clinical basis. Culture of the parasites is possible, but not really feasible in primitive rural conditions. Serology in espundia can be positive or negative (the quality of the antigen is of crucial importance). A practical problem in South America is whether a certain skin lesion with *Leishmania* amastigotes is caused by *L. braziliensis* or not. The geographical origin of the lesion or PCR and/or zymodeme analyses may give an answer here, though these laboratory techniques are not available in rural areas.

4.3.4   Differential Diagnosis

Differential diagnosis includes skin cancer, tertiary syphilis and yaws, leprosy, rhinoscleroma (a very chronic granulomatous infection with *Klebsiella rhinoscleromatis*),

rhinosporidiosis, midline granuloma (a form of T-cell lymphoma), Wegener's granu-
lomatosis, sarcoidosis, skin tuberculosis, infection with the free-living amoeba
Balamuthia mandrillaris, chronic nasal cocaine abuse, noma, and fungal infections
such cryptococcosis, histoplasmosis and South American blastomycosis (paracoc-
cidioidomycosis). With this last disease, which is a very chronic infection, the lungs
are frequently affected in a manner that can mimic tuberculosis. The yeast has typical
oval cells with ectospores and can be detected in sputum.

4.3.5   Treatment

Systemic drugs should be given. In South America meglumine antimonate® (20 mg
Sb/kg/day for 30 days) is first choice. Beware: the dose is expressed in mg antimony
(Sb). Alternatives are amphotericin B (Fungizone®), pentamidine and allopurinol.
Ketoconazole can be used in *L. mexicana* infections. Plastic surgery is sometimes
necessary in espundia in case of disfigurement. The use of pentoxyphylline (a xan-
thine derivative) plus antimony in refractory mucosal leishmaniasis is still in the
experimental phase. Pentoxyphylline 400 mg t.i.d. orally for 30 days should have an
anti-TNF-α effect. It is better known for its vasodilating effects (use in chronic
occlusive diseases of the legs).

4.3.6   Prevention

Control of leishmaniasis is sometimes restricted to these "minimal control initia-
tives" owing to, for example, financial reasons. Vector control with insecticides in
and around dwellings. This can be coordinated with malaria and/or Chagas' disease
control. Use of a fine gauze mosquito net impregnated with permetrine if transmission is taking place via a nocturnal biting vector. Biotope modification: environ-
mental sanitation in order to destroy the sandflies' breeding sites. An area of 300 m
radius is sometimes cleared of vegetation around villages.

Reservoir control: Sick dogs are treated or killed (*L. infantum*). Pets can wear a
dog-collar impregnated with deltametrine, permetrine of phenthione or be treated
with a lotion containing these active compounds. Rodents (*L. major*) are suitably
controlled by various methods: poison such as anticoagulants, deep ploughing to
destroy their holes and passageways, elimination of their breeding sites (rubbish
dumps) and of food plants (such as goosefoot [*Chenopodium*] for *Psammomys obe-
sus* ("fat sand rat"). Vaccination with *L. major* on aesthetically unimportant body
sites. This causes a lesion for a varying number of months which sometimes has to
be treated. New developments are vaccines based on killed *Leishmania* organisms
mixed with BCG and vaccines based on genetically modified *Leishmania* parasites
and variants of these. The value of vaccination remains disputed and vaccination is
not practised routinely.

4.4   Visceral Leishmaniasis (VL)

4.4.1   Diagnosis

VL should be suspected in case of persistent fever, emaciation, hepatosplenomegaly and signs of pancytopaenia. Differential diagnosis includes tuberculosis, brucellosis, Hodgkin's or non-Hodgkin's lymphoma, leukaemia, myelofibrosis, chronic malaria, liver cirrhosis, schistosomiasis with liver fibrosis and, more rarely, metabolic diseases such as Gaucher type 1 disease (glucocerebrosidase deficiency) or Nieman Pick type B disease (sphingomyelinase deficiency). Direct diagnosis is made by demonstrating the presence of the parasite. This is usually done on bone marrow obtained by sternum aspiration. The parasite is egg-shaped and measures $2–3 \times 5$ µm. There is a pale blue cytoplasm, a well defined nucleus and a smaller kinetoplast (Giemsa staining). In very rare instances, when there is a considerable delay between the aspiration itself and the preparation of the microscopic slide in the laboratory, promastigotes can be seen. In these cases the amastigotes had the time to change their morphology. The technique of spleen aspiration is more sensitive (in some studies very nearly 100 %, though in reality slightly lower) than bone marrow aspiration, but can be risky (spleen rupture, haemorrhage). If one wants to use a splenic aspirate, it is better to use the intercostal rather than the transabdominal route (safer, can be carried out more often). The 10th intercostal space between the anterior and the mid-axillary line is generally used. The procedure is safe when performed by an experienced physician, and when the prothrombin time is normal. The platelet count should be above $40 \times 10^9/l$. One can use a 21-gauge needle and a 5 ml syringe. After penetration of the skin, the plunger is withdrawn, the needle is quickly inserted into the spleen while maintaining suction and withdrawn immediately. Lymph node aspiration and liver biopsy are sometimes necessary. The parasites can rarely be detected in peripheral blood monocytes. Serology is positive in visceral leishmaniasis. The DAT (direct agglutination test) is often used, as this test has a high sensitivity and specificity. The best possibility in a small regional clinic is to absorb a drop of blood from a patient suspected to have kala azar on a small filter paper and then to punch out a standard size disk from the blood spot. In this way one obtains a well-defined, accurate aliquot of absorbed blood. This can be transported and used for DAT in a well equipped laboratory. Serology will remain positive after cure. If *Leishmania*-HIV co-infection is suspected, blood smears buffy coat or blood cultures can be used to detect the parasites. The parasites can be cultured, both in vitro (NNN-medium, Schneider's medium, Tobie rabbit blood agar (only Old World) or Nogushi soft agar with rabbit blood (Old and New World), as well as in test animals (golden hamsters), though this is usually not feasible in clinical practice in developing countries. A *Leishmania* parasite can survive for 3 days at a temperature of 4 °C, but for only 1 day at room temperature, in Locke transport medium (a buffered glucose-salt solution with antibiotics).

4.4.2 Treatment

Visceral leishmaniasis is currently treated with pentavalent antimony derivatives (antimony, chemical symbol Sb=Stibium). The derivative most frequently used is Glucantime® (meglumine antimonate, 85 mg Sb/ml) and, rarely, Pentostam® (sodium stibogluconate, 100 mg Sb/ml). The drugs can be administered IM (intramuscularly, painful) or by slow IV (intravenous) injection or infusion (diluted with 5 % glucose solution, otherwise local thrombophlebitis occurs). The dose is always expressed as mg Sb: 2×10 mg/kg IM or slow IV infusion per day for at least 30 days. As a dose is practically totally excreted and eliminated via the urine within 6 h after administration, a twice daily administration would pharmacokinetically be more logical than an injection once daily. However, a single administration per day appears to suffice in practice. The dose should be reduced in patients with kidney failure. A maximum of 850 mg/day [10 ml Glucantime®] has been previously set due to the risk of cardiotoxicity with higher doses. Negativation of T-waves and prolongation of the QT-time are indicative of threatening arrhythmia. The fever usually disappears after 1 week. The spleen begins to get smaller after 2 weeks but frequently requires 6–12 months to return to normal.

Laboratory diagnosis of leishmaniasis can be made by light microscopic examination of the stained specimen, by in vitro culture, by animal inoculation, detection of parasite DNA in tissue samples, detection of parasite antigen or specific antileishmanial antibodies (Sundar and Rai 2002). Direct visualization of amastigotes in clinical specimens is the diagnostic gold standard in regions where tissue aspiration is feasible and microscopy and technical skill are available. Diagnostic sensitivity for splenic, bone marrow, and lymph node aspirate smears is >95 %, 55–97 %, and 60 %, respectively (Sundar 2003). Elsewhere, serum antileishmanial immunoglobulin G in high titre is the diagnostic standard, primarily with direct agglutination tests or other laboratory-based serological assays (Herwaldt 1999; Desjeux 2004; Abdallah et al. 2004). Freeze-dried antigen and rapid detection of anti-K39 antibody with fingerstick blood in an immunochromatographic strip test have advanced field serodiagnosis. In symptomatic patients, anti-K39 strip-test sensitivity is 90–100 % (Sundar 2001a, b). This test can safely substitute for invasive diagnostic procedures in VL and is also useful for PKDL. Testing urine for leishmanial antigen or antibody is a new approach (Islam et al. 2004; Sundar et al. 2005). Various immunodiagnostic tests include antibody detection, complement fixation test, immunodiffusion test, CCIEP, indirect hemagglutination, IFA test, DAT, ELISA with CSA, ELISA with fucose-mannose ligand, ELISA with rK39 antigen and rapid strip test with rK39. In addition, antigen detection test called KATEX is 68–100 % sensitive. DNA detection by PCR with LDI primer using blood, bone marrow and skin are currently under use (Sundar et al. 2006). Different DNA sequences in the genome of *Leishmania* have been documented in diagnosis and prognosis of VL (Sundar et al. 2002). The drugs currently recommended for the treatment of VL include the Pentavalent Antimonials, Amphotericin B and its lipid formulations (AmBisome®), Pentamidine, Miltefosine, Paromomycin and Sitamquine.

4.5   Post Kala-azar Dermal Leishmaniasis (PKDL)

A skin condition, called post-kala azar dermal leishmaniasis (PKDL), can occur after a patient has suffered from kala azar. PKDL rarely occurs without being preceded by kala azar. This disease (PKDL) was originally described by Brachmachari in India. PKDL occurs on average 4 months after kala azar, though there are strong regional variations. This disease occurs mainly in India (up to 20 % of kala azar patients), and to a much lesser extent in the Middle East. In Sudan the disease occurs regularly (56 % of kala azar patients in one study). It is virtually unknown in the Mediterranean Basin or in South and Central America. It involves discoloured patches and painless nodules on the skin that usually contain few, but sometimes moderate numbers of amastigotes. Most of the lesions occur on the face (98 %) and to a lesser extent on the thorax (80 %), arms (70 %), legs (40 %), tongue (40 %) and genitals (6 %). Various degrees of severity can be clinically distinguished in PKDL. Grade 1 includes an extensive maculopapular to nodular rash, principally around the mouth, and possibly somewhat lesser lesions on the thorax and upper arms. Grade 2 is a similar but denser rash that covers the whole face and is also present on the chest, back, upper arms and legs, with fading of the lesions in more distal regions of the body. Grade 3 is a generalised dense rash with ulcers, scabs, cheilitis and possibly lesions of the palate (roof of the mouth). This disease has a very chronic course (years) and is therefore important for transmission. Parasites do not affect internal organs in PKDL. There is sometimes a concomitant neuritis, which can further contribute to the clinical resemblance to leprosy.

Chapter 5
Antimonials and Resistance

5.1 Antimonials and Leishmaniasis

Antimonials have been known since ancient times as medicine. It is called as antimonials because antimony (Sb)—a metalloids belonging to group XV of periodic table of element present in this compound. Sb is present in trivalent Sb(III) and pentavalent Sb(V) form. In the beginning of twentieth century, Tartar Emetic, an organic trivalent form of antimony was used for treatment of sleeping sickness and Gaspar Vianna introduced the drug for the treatment of MCL in 1912. The activity of Antimonials against various clinical form of leishmaniasis was soon discovered. Since, Tartar Emetic was highly toxic for human, new less toxic and more effective Antimonials were developed and evaluated. Pentavalent [Sb(V)] salts were found to be less toxic to mammal than trivalent [Sb(III)] salts. By the end of 1940s pharmaceutical research produced the current drug in use; the closely related organic pentavalent antimonial compounds Sodium antimony Gluconate (SAG) and Meglumine antimonite.

5.1.1 Leishmanicidal Activity of SAG

The intrinsic susceptibility of *leishmania* to SAG varies between different species. *In vitro* studies using MΦ-amastigote models have shown that *L. donovani* is three to fivefold more susceptible to SAG than *L. Major, L. tropica, L. maxicana*. Likewise susceptibility for SAG varies in different life stages of *leishmania*. Promastigote tolerate about 100-fold more concentration than intracellular amastigote. Reduced form of Sb(III) is highly toxic for both life stages of *leishmania*. This antimony patterns reflects hypothetic dual mode of action of SAG:

i. Upon contact with infected MΦ SAG helps the MΦ to kill its intracellular guest and when reaching the to the parasite, Sb(V) is reduced to Sb(III).
ii. Sb(III) can further directly kill the parasite.

A. Kumar, *Leishmania and Leishmaniasis*, SpringerBriefs in Immunology 3,
DOI 10.1007/978-1-4614-8869-9_5, © Springer Science+Business Media New York 2013

5.1.2   Interaction of SAG and Host Cell

Interaction between SAG and infected MΦ received little attention in *leishmania* studies on Antimonials, although there always have been clear evidence pointing out the importance of MΦ for SAG. Research's observations suggested that SAG requires MΦ in functional immune system to fully unfold its activity. As we know that *L. donovani* modulates signaling pathway of host cell extensively in favour of its own survival. Recent research now suggests that SAG modulates the signaling pathway of MΦ thereby restoring the MΦ's protective mechanism against the parasite.

5.1.3   SAG Triggers RNI/ROI Generation

In vitro treatment with 10 mg/ml SAG (comparable to drug concentration in blood) of uninfected MΦ induces the generation of both RNI and ROI. ROI production is immediate and already breaks at 3 h after treatment and fading out 3 h latter. This effect has also been observed in whole blood of animal after SAG treatment.

5.1.4   SAG Activates Interleukin (IL)

SAG treatments of infected or uninfected MΦ also induce expression of IL-12. It plays central role in development of specific T-cell response crucial for control of *Leishmania* infection. However in active VL, T-cell response is futile due to suppressive action of omnipresent IL-10. SAG treatment probably helps reversing this suppressive state by inducing IL-12 expression. Specific T-cell response acts mainly through IFN-γ which activates the infected MΦ but *L. donovani* paralyzes IFN-γ responsiveness of host cell. Thus SAG could restore IFN-γ responsiveness and thereby switching on antileishmanicidal activity of MΦ. In vitro study already demonstrates increased JAK/STAT phosphorylation (signaling downstream IFN-γ) upon SAG treatment. In addition the expression of IFN-γ receptor is also up regulated in response to SAG treatment which could further contribute to IFN-γ responsiveness of the MΦ.

5.1.5   Interaction of Sb(V)/Sb(III) with Leishmania

Sb(V) is relatively inert and exerts minimal direct toxic activity against *Leishmania*. However Sb(V) is metabolically converted *in vivo* or *in vitro* to Sb(III) which is toxic to both MΦ and amastigote, through its oxidative properties, binding functional thiol-group through its proteins and peptides. Sb(V) and Sb(III) both taken up by Promastigote and amastigote form of *Leishmania*. The uptake route of Sb(V) has not been defined clearly. Possibly, the uptake of Sb(V) is mediated by a transporter of *Leishmania* recognizing a sugar like structure resembling the gluconate like portion of SAG. Sb(III) enters inside cell through aquaglyceroporins (AQP) (Liu et al. 2002). AQP belongs to a major intrinsic membrane protein family that transports water and other uncharged polar solute such as glycerol and urea bidirectional. Several genes encoding AQP have been identified in the sequenced genome of kinetoplastids. One aquaglyceroporins (AQP1) has been partially characterized in *L. major* and *L. tarantole*.

5.1.6 Mechanism of Action of SAG

After 60 years of use, the antileishmanial mechanism of action of pentavalent antimonials [Sb(V)] is only now nearly understood. Nonetheless, it is now generally accepted that all pentavalent antimonials are prodrugs that require biological reduction to the trivalent form [Sb(III)] for antileishmanial activity. The site (amastigote or MΦ) and mechanism of reduction (enzymatic or non-enzymatic) remain controversial. Further studies are required to resolve this issue. Although stage-specific reduction has been demonstrated recently (Shaked-Mishan et al. 2001), the mechanism by which amastigotes reduce Sb(V) is not clear. Both glutathione and trypanothione can nonenzymatically reduce Sb(V) to Sb(III), particularly under acidic conditions. However, the physiological relevance of these observations is open to question since the rates of reduction are rather slow. Moreover, promastigotes contain higher intracellular concentrations of trypanothione and glutathione than amastigotes and both stages maintain intracellular pH values close to neutral, independent of external pH. Thus, it is difficult to account for the selective action of Sb(V) against the amastigote stage by a nonenzymatic mechanism. As both stages can take up Sb(III) and Sb(V) the insensitivity of promastigotes to Sb(V) cannot be attributed to drug exclusion (Brochu et al. 2003). Two possible candidates for the enzymatic reduction of Sb(V) to Sb(III) in amastigotes have recently been identified. The first is a thiol-dependent reductase related to glutathione S-transferases that is more highly expressed in amastigotes. The second is a homologue of a glutaredoxin-dependent yeast arsenate reductase.

The levels of expression of this protein in promastigotes and amastigotes were not reported and the low specific activity of the recombinant enzyme with glutaredoxin raises questions as to the physiological nature of the electron donor in *Leishmania spp.* The importance of these candidate proteins in conferring sensitivity to Sb(V) in amastigotes needs to be addressed. There have been comparatively few studies on the mode of action of these drugs. Initial studies suggested that sodium antimony gluconate (SAG) inhibits macromolecular biosynthesis in amastigotes (Berman and Gallalee 1987), possibly via perturbation of energy metabolism due to inhibition of glycolysis and fatty acid β-oxidation (Berman and Gallalee 1987). However, the specific targets in these pathways have not been identified. More recent studies have reported apoptosis in Sb(III)-treated amastigotes involving DNA fragmentation and externalization of phosphatidylserine on the outer surface of the plasma membrane (Sereno et al. 2001). However, these effects do not

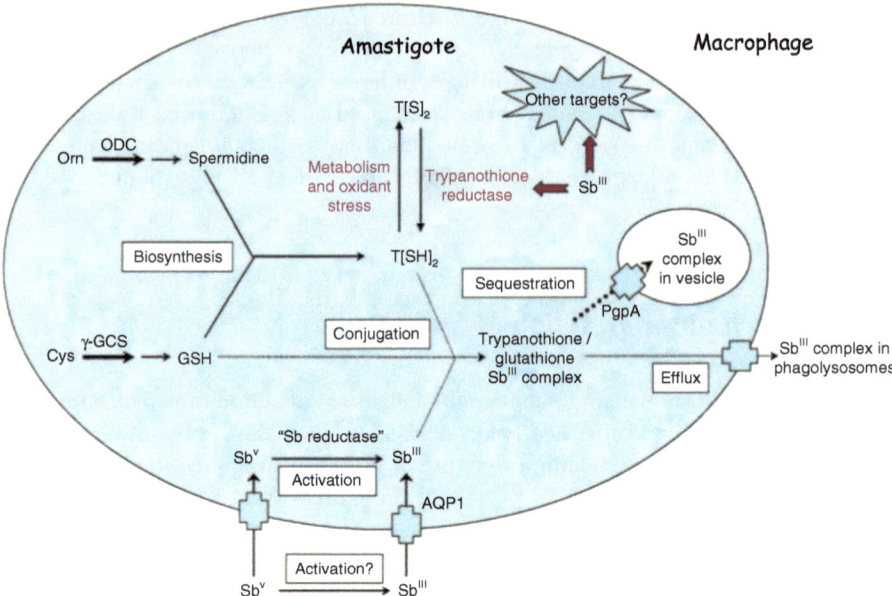

Fig. 5.1 Proposed mechanisms of antimony action and resistance in Leishmania spp. Levels of ornithine decarboxylase (ODC), γ-glutamylcysteine synthetase (GCS), and an intracellular P-glycoprotein (PgpA) are elevated in some laboratory-derived resistant lines (*thick lines*), whereas decreased Sb reductase is observed in others. *Dotted lines* indicate nonenzymatic steps implicated in resistance. The *red arrow* indicates inhibition of trypanothione reductase and other targets. Uptake of Sb(III) is mediated via an aquaglycoporin (AQP1) (Diagram courtesy of Croft et al. 2006)

involve the classical caspasemediated pathway (Sereno et al. 2001) and do not meet the more recent stringent definition of apoptosis.

The mode of action of antimony in drug-sensitive *L. donovani* involves several effects on glutathione and trypanothione metabolism (Fig. 5.1). Exposure to Sb(III) causes a rapid disappearance of trypanothione and glutathione from isolated amastigotes and promastigotes in vitro. A significant portion of these thiols are effluxed from cells in approximately equimolar amounts with the remainder being converted intracellularly to their respective disulfides (trypanothione and glutathione). The formation of the latter was ascribed to continuing oxidative metabolism in the face of inhibition of trypanothione reductase. Sb(III), but not Sb(V), has previously been shown to be a time-dependent reversible inhibitor of trypanothione reductase in vitro (Cunningham and Fairlamb 1995). Since Sb(III) also inhibits recovery of intracellular thiols following oxidation with diamide, this is consistent with inhibition of trypanothione reductase in intact cells. The profound loss of these thiols (>90 % in 4 h) coupled with the accumulation of disulfide (up to 50 % of the residual within 4 h) causes a marked decrease in cellular thiol redox potential. Similar effects on thiol levels and thiol redox potential were observed when amastigotes were exposed to Sb(V), intrinsically linking the effects of the biologically active Sb(III) with clinically prescribed Sb(V).

5.2   Antimonials Treatment, Failure and Resistance in *Leishmania*

Sb(V) has been highly effective in the treatment of Indian VL at a low dose (10 mg/kg) for short durations (6–10 days). But in the early 1980s, reports of its ineffectiveness emerged and the dose was eventually raised to 20 mg/kg for 30–40 days. In recent years, the proportion of patients unresponsive to Sb(V) has steadily increased. In hyperendemic districts of north Bihar, as many as 65 % of previously untreated patients fail to respond promptly or relapse after therapy with antimony drugs, due to the development of drug resistance (Sundar 2001a, b). The reason for the emergence of resistance is widespread misuse of the drug, as Sb(V) is freely available in India, and is easily accessible over the counter. In the endemic regions, unqualified doctors, irregular use of drugs and incomplete treatment were of common occurrence. These practices presumably expose the parasites to drug pressure, leading to progressive tolerance of parasite to Sb(V) (Sundar 2001a; Croft et al. 2006). The mechanism by which *Leishmania spp.* acquire resistance to antimonials has been the subject of intensive research for several decades, often yielding apparently contradictory results. It is also not inconceivable that some *Leishmania spp.* constitutively express higher amounts of "antimony reductase" activity in the promastigote stage than others. Diminished biological reduction of Sb(V) to Sb(III) has been demonstrated in *L. donovani* amastigotes resistant to sodium stibogluconate (Shaked-Mishan et al. 2001). This line also shows cross-resistance to other Sb(V) drugs, but the same susceptibility to Sb(III) as the wild type (Ephros et al. 1999), distinguishing it from the trypanothione pathway. It is not known whether this mechanism occurs in clinical isolates at present. The accumulation of Sb(V) and Sb(III) in promastigotes and amastigotes has been shown to be by different transport systems (Brochu et al. 2003), and although Sb accumulation was lower in resistant forms than in sensitive forms of parasite.

Aquaglycoporins have recently been demonstrated to mediate uptake of Sb(III) in *Leishmania spp.* and overexpression of aquaglycoporin 1 (AQP1) renders them hypersensitive to Sb(III) (Goyard et al. 2003). Transfection of AQP1 in an Sb(V) resistant field isolate also sensitized it to SAG when in the amastigote form in a macrophage. Increased levels of trypanothione have been observed in some lines selected for resistance to Sb(III) or arsenite (Mukhopadhyay et al. 1996). This is due to increased levels of the rate-limiting enzymes involved in the synthesis of glutathione (γ-glutamylcysteine synthetase) (Grondin et al. 1997) and polyamines (ornithine decarboxylase) (Haimeur et al. 1999), the two precursor metabolites to trypanothione. Increased synthesis of glutathione and trypanothione from cysteine could help to replace thiols lost due to efflux as well as to restore thiol redox potential perturbed by accumulation of disulfides (Wyllie et al. 2004). Spontaneous formation of Sb(III) complexed with either glutathione, trypanothione or both has been demonstrated by proton nuclear magnetic resonance spectroscopy (Yan et al. 2003) and by mass spectrometry (Mukhopadhyay et al. 1996). Since glutathione S-transferase (GST) is elevated in mammalian cells selected for resistance to arsenite, it has been proposed that formation of the metalloid-thiol pump substrates in *Leishmania spp.* could be rate-limiting and that GST could mediate this activity (Mukhopadhyay et al. 1996). However, GST is not detectable in *Leishmania spp.*, although there is an unusual trypanothione S-transferase activity associated with the eukaryotic elongation factor 1B complex (Vickers et al. 2004). The precise nature of the Sb-thiol complex remains uncertain, but two routes of elimination of the complex can be envisaged. The first involves sequestration in an intracellular compartment or direct efflux across the plasma membrane. Early studies noted that PgpA, a member of the ATP-binding cassette (ABC) transporters, is amplified in some resistant lines (Ouellette and Borst 1991). However, it soon became apparent that this transporter is not responsible for drug efflux across the plasma membrane. First, overexpression of PgpA was reported to decrease influx of Sb rather than increase efflux, possibly due to a dominant-negative effect through interactions with other membrane proteins. Second, overexpression of PgpA did not mediate increased efflux of radioactive arsenite from cells or transport of arsenite across plasma membrane preparations (Mukhopadhyay et al. 1996). Finally, PgpA plays a relatively minor role in resistance (Perez-Victoria et al. 2002) and is localized in membranes that are close to the flagellar pocket, the site of endocytosis and exocytosis in this parasite.

Chapter 6
Vaccines Against Leishmaniasis

The development of a vaccine against leishmaniasis is a long term goal in both human and veterinary medicine. In the past decade, various subunit and DNA antigens have been identified as potential vaccine candidates in experimental animals but none have so far been approved for human use. To date there is no vaccine available against VL in routine use anywhere in the world. However, there is consensus that in the long term, vaccines ought to become a major tool in the control of this group of diseases. Unfortunately, the development of vaccines has been hampered by significant antigenic diversity and the fact that the parasites have a digenetic life cycle. Although a great number of antigens have been tested for protection against the cutaneous disease with *in vitro* cell or mouse models, no effective vaccine against human kala-azar is yet available. Though, the solid immunity observed following cure of kala-azar has suggested that the vaccination to prevent leishmaniasis is within the reach of conventional immunization methods, only few reports in literature deal with vaccines viz., FML, FML-QuilA Saponin against canine VL (Santos et al. 2002). Immune mechanisms involved in VL and various vaccines candidates evaluated in experimental models for VL has been extensively reviewed by Modabber (1995), Handman (2001), Scott (2003), Ravindran and Ali (2004), Goto and Lindoso (2004), Kaye et al. (2004), Wilson et al. (2005), Garg and Dube (2006).

The ultimate goal of a vaccine is to develop long-lived immunological protection. Vaccination leads to enhanced responses that either completely prevent infection or greatly reduce the severity of the disease. Therefore, the important step in a rational design of a vaccine is to understand the immunological correlations of protection. Leishmaniasis in general, but particularly CL is probably one of a few parasitic diseases that are most likely to be controlled by vaccines. The relatively uncomplicated leishmanial life cycle and the fact that recovery from a primary infection renders the host resistant to subsequent infections indicate that a successful vaccine is feasible. The immunity observed after the therapeutic cure of kala-azar, suggests that vaccination against leishmaniasis is feasible and within the reach of conventional immunization methods. The advantages with leishmaniasis are

- Promastigote form can largely be grown in *in vitro* culture easily and are therefore amenable to manipulations required for vaccine development.
- The infection produces a state of premunition (concurrent infection and immunity) and this would explain the long lasting protection to reinfection often seen in recovered individuals. A number of reviews (Handman 1997) have summarized various aspects of vaccination attempts both in men and experimental animals.
- It is possible to induce resistance to reinfection in experimental animals by vaccination.

Several vaccine strategies have been developed. The "first generation" vaccines composed of killed parasites with or without adjuvant, are undergoing various stages of phase I (safety), II (reactivity) and III (efficacy) trials in humans conducted in Brazil. Vaccines with BCG as adjuvant have better protection as effective as chemotherapy. The "second generation" vaccine candidates live mutant vaccines, defined subunits and crude fractions are in preclinical development stages (Modabber 1995). Finally, protection results obtained with the third-generation vaccines composed of cDNA encoding leishmanial antigens cloned into a eukaryotic expression vector are still in a preliminary stage (Handman 1997). An ideal anti-leishmanial vaccine would need to possess several attributes, but not all of them may be easily achievable. These include (I) Safety; (II) Affordability and Stability; (III) A good vaccine should stimulate a strong, protective and long lasting immune response, (IV) Effectiveness against species causing CL and VL, (V) Stability at room temperature; (VI) A good vaccine should induce the appropriate immune responses- (VII) Effectiveness as a prophylactic as well as a therapeutic vaccine; (VIII) A single dose of vaccine should confer long-lived immunity.

6.1   Vaccination Strategy and Vaccines Against Leishmanisis

To date, there is no vaccine against *Leishmania* in routine use anywhere in the world. Several vaccine preparations are in more or less advanced stages of testing.

6.1.1   Live Vaccination

With the establishment of culture conditions that are able to support the growth of promastigotes Nicolle and Manceau in 1908, use of live organisms to control infections was started). Leishmanization, the inoculation of live virulent *Leishmania*, has been practiced for over a century (Modabber 1995). The injection of viable parasites produces presumably controlled lesion and induces T cell mediated immunity. Leishmanization was used successfully for a long time in the republic of the former

Soviet Union, Israel and Iran which was however, abandoned in most countries mainly due to safety issues (Modabber 1995). Efforts are being made to improve safety of leishmanization by the inclusion of drug-sensitive *Leishmania* mutants with suicide genes for controlled infection, inclusion of killed parasites to reduce the size and duration of lesions, or by using adjuvants that promote more rapid onset of anti-leishmanial immunity and swift healing of lesions. Amongst the earliest vaccination strategies attempted in hamsters, was the use of live amastigotes, which failed to impart protection against challenge with *L. donovani*. Even an effort to use BCG as adjuvant, a strategy that induced some level of resistance in mice, led to an increase in susceptibility when tried in hamsters. An approach using live promastigotes of UR6 strain of *L. donovani* could, however, impart significant protection in both spleen and liver (Russell and Alexander 1988).

6.1.2 Live-Attenuated Vaccines

Vaccination with attenuated parasites, which are infectious but not pathogenic, has major advantages compared to Leishmanization or vaccination with killed promastigotes. Attenuated parasites are taken up by the natural host cell into the same compartment as the virulent organisms and persist long enough for the induction of the appropriate immune response without causing disease. Attenuated *Leishmania* vaccines have been produced by long-term culture, by culturing under gentamycin pressure, irradiation, chemical mutagenesis and, more recently, by deleting genes from the *Leishmania* genome. Parasites lacking genes essential for long-term survival in the mammalian host, such as the gene encoding dihydrofolate reductase-thymidylate synthetase (DHFR-TS) were tested as potential vaccines. However, vaccination with *L. major* lacking DHFRTS failed to show protective immunity in a primate model (Amaral et al. 2002). The limited protection conferred by these attenuated parasites might be due to their rapid elimination by the host. Targeted deletion of other virulence genes, such as the cysteine proteinase genes or the lpg1 gene, encoding a putative galactofuranosyl transferase involved in the biosynthesis of the virulence factor, lipophosphoglycan, resulted in parasites which, although attenuated, could still cause disease, making such parasite lines unacceptable as vaccines (Ryan et al. 1993; Alexander 1988). Recently, *L. major* parasites lacking the lpg2 gene encoding an enzyme involved in the transport of GDP-mannose to the Golgi apparatus, were shown to persist in BALB/c mice without causing lesions, and to protect against homologous infection. Surprisingly, the protection induced by these attenuated organisms was not associated with a significant Th1 response as measured by IFN-γ production and delayed type hypersensitivity (DTH), suggesting that the Th1-like responses might not always be essential or correlate with protective immunity.

In summary, the use of attenuated organisms is very attractive because they are the closest mimic to the natural course of infection and may therefore lead to similar immune responses. Moreover, because of the small load of antigen delivered by the transient infection, the immune responses may be skewed even more toward a Th1 protective response than in natural infection (Handman 2001). Such immunization will also deliver many more parasite antigens than the limited number possible with subunit or recombinant antigens. Summarizing a large amount of experimental evidence, concluded that injection of attenuated organisms achieved better protection than any method involving recombinant gp63 as test antigen delivered with a variety of adjuvants and delivery systems. If this conclusion is shown to be generally applicable to other vaccine candidates, the prospect of using attenuated *Leishmania* vaccines in preference to subunit or recombinant approaches will gain favor. The disadvantages of such vaccines are the logistics of their large-scale production and distribution in the field.

6.1.3 Killed Vaccines

Killed parasite vaccines have been proposed as both prophylactic and therapeutic vaccines. Several strategies including the use of attenuated or killed parasites have been tried against murine VL. A number of first-generation killed vaccines have entered vaccine trials against cutaneous leishmaniasis giving encouraging results. Among the first attempts was the use of killed promastigotes together with glucan as an adjuvant injected subcutaneously or intravenously which could impart partial resistance against infection in CF1 mice. Vaccination studies by Mayrink and colleagues in new areas confirmed previous reports (Mayrink et al. 1979, 1985). In another vaccination studies promastigotes of five killed *Leishmania* strains were shown to be safe and immunogenic as measured by the leishmanin skin test (LST) conversion, but conferred only a small degree of protection (50 %). The production of a vaccine known as Leishvacin containing only one of the initial strains (*L. amazonensis*, IFLA/BR/67/PH8) was subsequently initiated. It was suggested, after vaccination studies in mice as well as in vervet monkeys, that Th1 immune response may be necessary for protection against infection, but not sufficient for protection against CL (Gicheru et al. 2001).

Convit and colleagues pioneered the use of immunotherapy with a combination of killed *L. amazonensis* promastigotes and BCG for the treatment of localized CL. In Venezuela, a clinical healing rate of more than 95 % was achieved and cure was associated with a Th1-like immune response in the patients. A modified form of the vaccine using pasteurized *L. braziliensis* promastigotes and live BCG was effective in the treatment of refractory mucocutaneous leishmaniasis and early cases of diffuse CL. In Brazil, the combination of killed *L. amazonensis* promastigotes with a half-dose regimen of meglumine antimoniate was also shown to be highly effective for the treatment of CL. In yet another study in Eastern Sudan it was observed that previous exposure with *L. major* appeared to protect the inhabitants against VL caused by *L. donovani*. Clinical trials involving the ALM vaccine regimen (autoclaved *L. major* together with BCG) have now entered phase III trials against cutaneous leishmaniasis in many parts of the world. The same preparation has also undergone randomized double blind trials against VL in an endemic region of

Sudan. Autoclaved *L. major* in combination with BCG was used and shown to be safe, and significant vaccine, it induced IFN-γ production in healthy volunteers. Alum precipitation of *L. major* improved the immunogenicity of the vaccine and induced a strong delayed type hypersensitivity (DTH) reaction in volunteers. Investigations are also made in the field of VL. Heterologous protection has been found using killed *L. major* plus BCG combination against *L. donovani* in Indian langur as well as in hamsters. This vaccine is under clinical trial in Sudan for prevention of VL. Similarly successful protection against *L. donovani* infection has been achieved with killed *L. major* in canine and hamster models and provides the basis for further human trials (Garg and Dube 2006).

6.1.4   Fractionated/Subunit Vaccines

With the exception of one single trial in humans, vaccination using *Leishmania* fractions has been confined to animal models. In animal models, subcutaneous injection of leishmanial antigens plus adjuvants (usually complete Freund's adjuvant; CFA) produce exacerbation or even blockage of the otherwise protective immunization which has been a major concern in human studies. Some workers also used *L. donovani* antigen from promastigotes, to characterize and discuses their possible use in serodiagnosis and immunoprophylaxis (Jaffe and McMahon-Pratt 1987; Reed et al. 1987). While major trials have been carried out with promastigotes, some workers have emphasized that the protective response is presumably generated by the intracellular amastigote form.

   Liposomised formulation of surface molecule glycoprotein gp63 and LPG from *L. mexicana* when inoculated in mice protected them to subsequent homologous challenge (Russell and Alexander 1988). Some leishmanial antigens extracted from the membranes of *L. donovani* promastigotes alone or in association with positively charged as well as negatively charged liposomes conferred high level of protection against infection with virulent promastigotes. Gp63 has been reported to be a major component of the Leishvacin vaccine tested in healthy volunteers in Brazil. Synthetic peptides as well as gp63 itself were used to pulse dendritic cells (DC) in order to investigate the potential of a DC-based subunit vaccine. Immunization with native promastigote PSA-2 protected mice against *L. major* infection (Handman et al. 1995). Protection of susceptible BALB/c mice by vaccination with secreted/excreted *L. major* antigens was attributed to a combination of PSA-2 and LPG contained in the preparation (Tonui et al. 2004). Amastigote protein A2 when used as a vaccine in either protein or DNA form against murine VL could significantly reduce parasites in the spleen (Ghose et al. 1980; Ghosh et al. 1995).

   Because of its immunopathogenic role in murine *L. major* infection, considerable interest has been focused on LACK (leishmaniasis homolog of receptors for activated C kinase) antigen as a potential vaccine candidate for leishmaniasis. However, vaccine strategies involving LACK against other species like *L. mexicana* and *L. donovani* were unsuccessful (Melby et al. 2001) though immunization of mice with a LACK, delivered in DNA form, could induce a robust parasite-specific Th1 immune response. A heterogeneous family of acidic surface molecules (named HASPs) expressed only in metacyclic and amastigote stages of the *Leishmania* life cycle was investigated as a potent vaccine candidate against murine VL. Different approaches have been taken in order to identify the most relevant molecules to be used in vaccination studies. One such approach consists of defining antigens recognized by PBMC or sera from patients or animals recovered from *Leishmania* infections. The PapLe22 antigen is recognized by T cells from VL patients and its vaccination led to a marked decrease in parasite burden in immunized hamsters. The leishmanial antigen ORFF was able to induce protective immunity against *L. donovani* combination with CpGs oligonucleotides (Ghosh et al. 1995). Molecules identified in this way include nuclear proteins (Suffia et al. 2000), promastigote surface protease gp63 (Kemp et al. 1993) (although a response to this molecule was not always observed (Jaffe et al. 1990), A2 protein of *L. donovani*, protein P-4 (a *L. pifanoi* membrane associated protein with nuclease and phospho-monoeslerase activity, LCR1 (a presumed flagellar antigen, LCR1 combination with BCG, cysteine proteinases, glucose regulated protein 78 (GRP78) of *L. donovani* (Jensen et al. 2001) and many other molecules of unknown function (Kemp et al. 1998; Suffia et al. 2000). T cells were derived from a donor residing in an area endemic for VL due to *L. chagasi*, whose skin test was positive without having developed an infection, suggesting that he had been able to develop a strongly protective immune response.

The low success rate of subunit vaccines was due to the fact that some of the polypeptides may be minor immunogens and so even though they may be excellent in a cocktail vaccine, individually they may provide only partial protection. For e.g. Recombinant Q protein formed by fusion of antigenic determinants from four cytoplasmic proteins from *L. infantum* (Lip2a, Lip2b, P0 and histone H2A) co-administered with live BCG protected 90 % of immunized dogs by enhancing parasite clearance. Leish-111f (Single poly protein composed of following three molecules fused in tandem TSA, LmSTI1 and LeIF all these three proteins were used as single in combination has been engineered and assessed in Phase I clinical trials in healthy volunteers. Early studies on vaccine development indicated that glycolipids such as the *Leishmania* lipophosphoglycan (LPG) provided excellent protection (Handman and Jarvis 1985; Russell and Alexander 1988). Protection depended on the use of adjuvants such as liposomes or *Corynebacterium parvum* and on the integrity of the molecule. Studies with *L. donovani*-derived LPG suggested that it might be a promising candidate for development of a vaccine. However, immunization with *L. donovani* LPG showed that experimental animals were not protected against homologous or heterologous challenge despite Th1 immune responses being observed.

6.1.5 Naked DNA Vaccines

Immunization with naked DNA is a new approach, which promises to revolutionize the prevention and treatment of infectious diseases. The gene encoding the vaccine candidate is cloned in a mammalian expression vector, and the DNA is injected directly into muscle or skin. Surprisingly, the plasmid DNA is taken up by cells and translocated to the nucleus, where it is transcribed into RNA and then translated in the cytoplasm. The efficiency of uptake and the expression of plasmid DNA must be extremely low, but there is abundant evidence that it is sufficient to provoke immune responses in both T and B cells. A large body of literature indicates that both CD4$^+$and CD8$^+$ cell mediated responses are induced, making a DNA vaccine attractive for a *Leishmania* vaccine. In addition of being able to induce the appropriate immune responses, DNA vaccines are attractive because they ensure appropriate folding of the polypeptide, produce the antigen over long periods, and do not require adjuvants. Another advantage is that the technology for production is very simple. DNA is stable, has a long shelf life, and does not require a strict cold chain for distribution. Vaccination with DNA encoding gp63, LACK, and PSA-2, cysteine proteinase type I (Cpb) and type II (Cpa) all protected both genetically resistant and susceptible mice from infection with *L. major*. Immunization with DNA encoding *L. infantum* protein paple 22 decreases the frequency of parasitemic episodes in infected hamsters (Fragaki et al. 2001). In a recent approach immunization with plasmid DNA isolated and pooled from 15 cDNA sub libraries (approx. 2,000 cDNAs/sublibrary) either collectively or singly conferred protection against parasite challenge in a murine model of VL (Melby et al. 2001).

A highly sophisticated approach for the identification of relevant antigens relies on the identification of peptides eluted from major histocompatibility class II molecules of infected antigen-presenting cells. A peptide obtained in this way from *L. donovani* including mouse macrophages was used to design an oligonucleotides primer; PCR performed on *L. donovani* DNA yielded a fragment which was cloned and used to screen an *L donovani* cDNA library. The corresponding 23 kDa protein (Ldp23) is expressed in both promastigotes and amastigotes of various *Leishmania* spp. In addition, T cells from mice immunized with *L. donovani* or infected with *L. major* or *L. amazonensis* responded to Ldp23 by producing high levels of IFN-γ and no IL-4, further stressing the potential of this molecule as a candidate vaccine. Several other antigens have been successfully tested as DNA vaccines against cutaneous or visceral infection. The former group include acidic ribosomal protein P0, P4 nuclease and paraflagellar rod protein 2 (PRP-2) (Saravia et al. 2005), whereas the latter contains ORFF (Sukumaran et al. 2003; Tewary et al. 2005), kinetoplastid membrane protein-11 (KMP-11), CPA and CPB and NH36, a main component of the fucose-mannose ligand (Basu et al. 2005).

6.1.6   *Recombinant and Synthetic Vaccines*

The newer vaccines under consideration comprise recombinant DNA-derived antigens and peptides. Some of the target antigens are species and life cycle stage specific, while others are shared by promastigotes and amastigotes. Some are conserved among *Leishmania* species, while others are not. Since T cells recognize peptides derived from cytosolic proteins bound in the MHC class I groove or peptides derived from the lysosomal compartment bound in the MHC class II groove on the antigen-presenting cell surface, it would appear that virtually any parasite protein might function as an antigen, regardless of its location in the parasite. Recombinant antigens can be delivered as purified proteins, as the naked DNA encoding them, or as bacteria manufacturing the proteins in situ. Manipulations now allow targeting of the antigen to specific locations or to particular antigen-presenting cells, such as dendritic cells or Langerhans cells, which are considered essential for the initiation of primary T-cell responses. Injection of bacteria or naked DNA may have the added advantage of providing an adjuvant effect, which may "activate" or "licence" these antigen-presenting cells.

(i) **Expression of Immunogens in Bacteria and Viruses**

The first recombinant antigen used to vaccinate against leishmaniasis was leishmaniolysin or gp63. This is a Mr 65,000 membrane protease present in promastigotes of all species. gp63 is one of the parasite receptors for host macrophages, and parasite mutants lacking the protein are avirulent. gp63 belongs to a multigene family, with different members being expressed in promastigotes and amastigotes. Interestingly, both the recombinant and native proteins seem to protect better against infection with *L. amazonensis* than against infection with *L. major*, suggesting species-specific epitopes, at least in animal models. It is unfortunate that in humans and animal models the T-cell responses to gp63 have been variable. However, when detected, they appeared to be of the Th1 type. Overall, gp63 is still considered a promising vaccine candidate. The gene has been engineered in a number of delivery systems (BCG, vaccinia virus, and *S. enterica* serovar Typhimurium) in the hope of inducing the appropriate Th1 immune response.

A second vaccine candidate tested in animal models is a membrane antigen of unknown function, gp46/M2 or parasite surface antigen 2 (PSA-2) (Handman 1997). Similar but distinct gene products are found in amastigotes and promastigotes of *L. major* and *L. donovani*, but in *L. mexicana* expression seems to be restricted to promastigotes. Recombinant DNA-derived PSA-2 protein was variable in its ability to confer protection, while the protein derived from the yeast *Pichia pastoris* provided good protection. These data suggested that the native conformation of the protein might be important for processing and presentation by antigen-presenting cells. These difficulties may be overcome by the development of a DNA-based vaccine. The leishmanial eukaryotic ribosomal protein (LeIF), a homologue of the ribosomal protein cIF4A, is being considered as a vaccine candidate based on its ability to induce Th1-type cytokines in humans (Probst et al. 1997). This protein is highly conserved in evolution, but assuming that specific parasite epitopes will be used for vaccination such that autoimmune responses will be avoided. A similarly conserved antigen, the *Leishmania* homologue of the receptor for activated C kinase (LACK), which is expressed by both promastigotes and amastigotes, has been shown to protect mice from infection, in particular when administered with IL-12 as an adjuvant. Interestingly, LACK is also the major target for Th2 responses in susceptible BALB/c mice, and BALB/c mice made tolerant to LACK are resistant to infection. Significance of this finding for the use of LACK as a vaccine in humans remains to be elucidated.

  L. Major recombinant antigens thiol specific antioxidant protein when immunized in Balb/c mice with or without IL-12 develops strong cellular immune response against *L. major* challenge. BCG expressing LCR1 of *L. chagasi* has been reported to induce protective immunity in susceptible mice. A stage specific hydrophilic acylated surface protein B1 confers protection against experimental visceral challenge. Eight other leishmanial antigens viz. flageller proteins α and β tubulin, histone, malate dehydrogenase, and elongation factor 2, as well as the two novel parasite proteins cloned, using parasite specific T cell lines derived from an immune donor, has been reported as T-cell stimulating antigens (Probst et al. 1997). Some are amastigote specific, such as A2, P4, and P8 of *L. mexicana pifanoi*. In view of the fact that the target of host protection is the amastigote, which has only a rudimentary flagellum, the mechanism which host achieves protection with this antigen is not obvious. A most interesting approach to the identification of potential vaccine candidates has been the elution of antigenic peptides from antigen-presenting cells (Campos et al. 1995). Several peptides were identified, and the sequences were used to clone the cognate genes. One of these genes encodes a membrane polypeptide expressed in promastigote and amastigote induced Th1-type responses in mice (Bunn-Moreno et al. 1985).

(ii) **Synthetic Peptides**

The 1980s were marked by a wave of enthusiasm concerning the use of peptide vaccines, in particular those considered to be T-cell epitopes. Several considerations make the peptide antigens less attractive: the magnitude of the T-cell memory induced the inability of all individuals in the population to respond to the peptide, and the economics of production. Since the antigenic peptide is processed and presented to T cells in the context of MHC class I or class II and since not all peptides associate with all MHC types, some peptides will not be recognized by all individuals in the population. There are additional "holes" in the ability to respond to individual peptides due to failure of processing, cleavage, transport or due to deletion of certain T-cell specificities due to self-tolerance. Despite these caveats, several *Leishmania* gp63 peptides have been tested successfully in animal models. Importantly, host protection was long-lasting, indicating the induction of long-term T-cell memory. Some polypeptides, such as PSA-2, need to be in their native conformation for antigen processing, and *Escherichia coli*-derived recombinant proteins may not fulfill this requirement. This problem may be overcome by exploitation of the parasites themselves by over expression of parasite antigens in transfected nonpathogenic *Leishmania* strains or the related trypanosomatid *Crithidia* (Kelley 1997). Presumably, polypeptides expressed in these systems will be abundant, correctly folded, and glycosylated (Pfeffer et al. 1990). Another reason for the low success rate of subunit vaccines is that some polypeptides may be minor immunogens and so even though they may be excellent in a cocktail vaccine, individually they may provide only partial protection. The immune responses in leishmaniasis can range from protective to positively harmful. These differences in the quality of the response are at least partly due to predominance of Th1 or Th2 cytokines and may be greatly influenced by antigen dose. Accordingly, the amount of antigen and possibly the route of its administration may be important issues. Another thorny issue concerns adjuvants.

6.1.7   Non-protein Antigens as vaccine

Early studies on vaccine development indicated that glycolipids such as the *Leishmania* lipophosphoglycan (LPG) provided excellent protection. Protection depended on the use of adjuvants such as liposomes or *Corynebacterium parvum* and on the integrity of the molecule. Not only was the water-soluble form of LPG lacking the glycosylphosphatidylinositol anchor not protective, but also it exacerbated disease. At the time when that work was published, the immune mechanism leading to host protection by such a nonprotein molecule was totally mysterious. Immunity was known to be T-cell mediated but T cells were not thought to recognize or present nonprotein antigens. Today, it is accepted that many novel and interesting microbial antigens including mycobacterial glycolipids can be recognized by T cells and that these antigens are presented to T cells by a special subset of MHC class I proteins known as CD1. In this context, it may be rewarding to reevaluate the potential of LPG as a vaccine candidate.

6.1.8   Anti-sandfly Saliva Components as Vaccine

Leishmania parasites are transmitted from one host to another during the sandfly bite as a suspension in sandfly saliva. Therefore, triggering immune responses against saliva components may indirectly enhance anti-leishmanial immunity. Early studies indicated that molecules in the sandfly saliva exacerbated the development of disease by facilitating the establishment of infection. On the other hand, prior exposure of mice to bites of uninfected sandflies seemed to confer protection from *L. major* infection. Protection was associated with a DTH response at the site of parasite injection (Kamhawi et al. 2000). Immunisation with molecules present in saliva, such as maxadilan (Morris et al. 2001) or a 15 kDa protein, SP15 (Valenzuela et al. 2001) also induced protection against CL. The role of saliva molecules in the natural infection process is not understood; neither the mechanism by which immune responses against them induces protection. Protective immunity can be directed towards immunomodulators in the saliva neutralising their ability to facilitate infection. Alternatively, the DTH response observed in all the studies may modify the injection site environment and render it inhospitable for *Leishmania* (Belkaid et al. 2000). Anti-saliva vaccination opens an interesting avenue in *Leishmania* vaccinology, and it is likely that such a vaccine would have additive effects when administered together with an anti-*Leishmania* vaccine.

6.2 Adjuvants

Adjuvants, a Latin word meaning, "to help," have been used to improve vaccine efficacy from the early 1920s. While the number and type of adjuvants have expanded, their mechanism of action has remained largely mysterious and empirical. Precisely, how adjuvants augment the immune response is not known, but they appear to exert different effects to improve the immune response to vaccine antigens, as such they (a) Improve antigen delivery to APCs, (b) Promote the activation state of APCs, (c) Enhance antigen processing and presentation by APCs, (d) Modulate antibody avidity and (e) Stimulate CMI. Alum, one of the few adjuvants to be licensed for human use was used together with killed parasites and exhibited significant resistance to subsequent infection with *L. donovani*. An alum-precipitated *L. major* vaccine was used in a study in combination with BCG in a single dose schedule against *L. donovani* in the langur model yielding moderate success. *Mycobacterium bovis* BCG, known to possess immunopotentiating properties, has been used as a viable vaccine against tuberculosis since the 1920s. An intravenous inoculation of BCG was found to be both prophylactic and therapeutic in BALB/c mice against challenge with amastigotes of *L. donovani*. Mice inoculated with BCG exhibited significantly lower levels of parasite burden when compared to controls. A similar attempt in hamsters against VL, however, exacerbated the disease and failed to protect the animals. Some other immunopotentiator including levamisole (Rezai et al. 1978), cyclosporin A (Bogdan et al. 1989), C. parvum (Hill 1987), glucan (Cook and Holbrook 1983), derivatives of MDP (Pal et al. 1991; Zehra et al. 1995), and saponins (Santos et al. 2002) are largely in use due to their ability to nonspecifically activate macrophages. Several other types of particulate adjuvants or delivery systems that have been used with leishmanial vaccines with variable outcomes include liposomes, microparticles, immunostimulating complexes, and micelles formed by intrinsically adjuvanted lipopeptides. One of the most promising adjuvants involves the use of some of "nature's adjuvants" i.e., soluble cytokines which are known to promote Th1 immune responses. Among these, IL-12 is essential for the induction and maintenance of Th1 immune responses by *Leishmania* vaccines, including DNA vaccines. IL-12 is a key component in the early response to *L. donovani* and neutralization of IL-12 over the first few days post infection leads to elevated parasite burdens in both the spleen and liver. While lL-12 was found to be effective adjuvant against CL, its use in a mouse model of *L. donovani* was found to be ineffective (Melby et al. 2001). Another interesting approach to the targeting of the immune response has been the coupling of the potent antigen-presenting ability of dendritic cells to the delivery of proinflammatory cytokines to the local site of response. For this purpose, adoptive transfer of dendritic cells engineered by retroviral infection to secrete IL-12 has been used to augment the effect of vaccination with dendritic cells pulsed with *L. donovani* antigen. Certain sequences of bacterial DNA seem to have immunostimulatory effects. Specific DNA sequences containing unmethylated dinucleotides (CpG motifs) activate B cells and dendritic cells and induce cytokine production by macrophages. The ability of CpG motifs to induce the production of IL-12 and TNF-α which lead to a polarized Th1 response makes them particularly attractive as adjuvants for *Leishmania* vaccines. CpG-containing DNA has immunostimulatory effects in vaccination.

6.2.1   Delivery of Adjuvants

Adjuvants, from the Latin word meaning, "to help," have been used to improve vaccine efficacy from the early 1920s. While the number and type of adjuvants have expanded, their mechanism of action has remained largely mysterious and empirical. Recent work from many laboratories has changed our thinking about the cells and molecules, which initiate immune responses. The current concept is that specialized "professional" antigen-presenting cells must initiate primary T-cell responses. These cells are the dendritic cells or the Langerhans cells in the skin. Dendritic cells via receptor-mediated endocytosis or fluid-phase pinocytosis take up antigen. Then migrate out of the tissue into the draining lymph nodes, where the antigen is processed into peptides and where the peptides meet the MHC class II molecules as they are assembled and transported through the endoplasmic reticulum and the Golgi apparatus. The complex of peptide with MHC class II is transported to the cell surface and displayed for recognition by T cells. The antigen-presenting cells also secrete cytokines, which attract CD4+T cells to the area, such that T cells with cognate receptors for the peptide MHC class II complex undergo clonal expansion. It is likely that the Th1/Th2 switch will occur at this point, thus determining the course and outcome of the immune response.

It is at this level that adjuvants are believed to contribute to the amplitude of the immune response and to its quality. The ability of bacteria and other particles to be taken up by dendritic cells, coupled with the ability to express foreign genes in bacteria, has made them attractive delivery vehicles for vaccines. Such a vaccine could exploit attenuated bacteria such as *Salmonella* or BCG, which are already in use as vaccines with demonstrated safety and immunogenicity in their own right. BCG has been used successfully for anti-*Leishmania* immunotherapy in South American patients without side effects. BCG vectors carrying gp63 have also been used successfully to induce protection in the *L. major* system. Another combination of alum-ALM+BCG has been used for VL in Indian langur monkeys. *S. enterica* serovar Typhimurium expressing gp63 has also been used in the mouse model. Unfortunately, the vaccine induced variable protection, ranging from minimal to significant, despite the induction of apparently appropriate T-cell responses (Yan et al. 2003).

6.2.2 Use of Cytokines as Adjuvant

One of the most promising adjuvants involves the use of some of "nature's adjuvants" i.e., soluble cytokines which are known to promote Th1 immune responses. Among these, IL-12 is essential for the induction and maintenance of Th1 immune responses by *Leishmania* vaccines, including DNA vaccines (Scott 2003). It appears that the persistence of IL-12 delivered, as DNA may be an important contributor to the long-term memory induced by the vaccinating DNA-encoded LACK antigen. An interesting approach to the targeting of the immune response has been the coupling of the potent antigen-presenting ability of dendritic cells to the delivery of proinflammatory cytokines to the local site of response. For this purpose, adoptive transfer of dendritic cells engineered by retroviral infection to secrete IL-12 has been used to augment the effect of vaccination with dendritic cells pulsed with *L. donovani* antigen.

Chapter 7
Experimental Models for Leishmaniasis

Murine models for experimental leishmaniasis are well established models. Parasites are injected underneath the skin of the footpad. While most of the mice strains like C57BL/6, CBA/J, C3H or BIOD2 resist the infection with clinical cure within few months (Handman et al. 1979); BALB/c and all T-cell immunodeficient strains manifest a systemic visceral leishmania leading to death (Howard et al. 1980, 1984; Mitchell et al. 1984). Resistance and susceptibility are closely related with the development of T-cell responses of Th1 or Th2 type, respectively. Recent studies have shown that the mice model of *L. donovani* does not reproduce the features of active human VL like chronic fever, hepatosplenomegaly, pancytopenia and profound cachexia and have an ineffective anti-leishmanial cellular response (Melby et al. 2001). On the contrary, Syrian golden hamster model of active VL closely relate the human counterpart as shown by relentless increase in visceral parasite burden, progressive cachexia, hepatosplenomegaly, pancytopenia, hypergammaglobulinemia and ultimately death (Melby et al. 2001). Unfortunately, studies in hamster model are limited by the lack of immunological reagents.

Success of any research work depends largely on the use of appropriate animal models and reliable techniques. For any kind of experimental work, the availability of these experimental models is desirable, as they provide valuable data on the efficacy of newly developed vaccines and immunotherapeutic agents against leishmaniasis. For immunoprophylaxis, availability of suitable animal models having physiological, pathological and immunological responses similar to human kala-azar is of prime importance. Animal models are expected to mimic the pathological features and immunological responses observed in humans when exposed to a variety of *Leishmania spp.* with different pathogenic characteristics. Many experimental models have been developed, each with specific features, but none accurately reproduces what happens in humans. For *in vivo* testing of vaccine several animal species have served as experimental host for VL. Important among them are BALB/c

Table 7.1 Experimental models for vaccine and chemotherapy research against experimental VL

Animal species	Scientific name	Immune response to	Advantages	Drawbacks
Mouse BALB/c C57BL/6	*Mus musculus*	Susceptible Resistance	Both inbred and outbred strains, immunodeficient available, Easy availability of immunological reagents	Does not shows progressive kala-azar symptoms
Syrian golden hamster	*Mesocricetus auratus*	Susceptible	Shows typical clinicopatho-logical profile similar to human cases	Lack of immuno-logical reagents limits studies
Dog	*Canis* spp.	Susceptible	Good model for canine studies and suitable for secandory screening	Difficulties in breeding and keeping
Indian langur	*Presbytis entellus*	Susceptible	Good model for tertiary screening and vaccine testing	Difficulties in breeding and keeping

mice and Syrian golden hamster (primary tests), dogs (secondary tests) and monkeys viz., squirrel, vervet and Indian langur monkeys as tertiary screens (Melby et al. 2001). Amongst these, hamsters and mouse and few non-human primates such as owl monkeys; *Aotus irivirgatus*, squirrel monkey; *Saimiri sciureus* and Indian langur; *Presbytis entellus* have been suggested as potentially useful model for various studies on VL (Garg and Dube 2006).

Amongst the above said two rodent models, murine model of *L. donovani* infection is a good model of early parasite replication followed by immunological control and sub clinical infection, but there is no murine model for the progressive disease observed in human active VL. Hamster develops symptoms and immunological consequences similar to human VL, resulting in a relentless increase in parasite burden, progressive cachexia, hepatosplenomegaly, hypergammaglobulinemia, pancytopenia, and ultimately death (Melby et al. 2001). However, immunological studies in this model are limited due to the lack of available immunological reagents. Therefore, the cytokine profiles of experimental hamsters have been standardized and carried out by RT-PCR in this study. Thus, they hold promise as primary model for immunotherapeutic/prophylactic trials. *Leishmania* cause cutaneous, mucocutaneous and visceral diseases in man depending on the species of the parasite and the host immune response. While extensive information is available about the immune response in experimental CL, the nature of immunity in experimental VL, which is different in many aspects, is poorly understood. In order to develop vaccines for different forms of leishmaniasis and since there are many areas where different species and different forms of the disease overlap, a detailed knowledge, particularity of the immune response and pathogenesis is extremely important. Although voluminous work has been carried out on development of vaccine against *Leishmania* a brief account on the various model adopted for development of *Leishmania* vaccines are described below (Table 7.1).

7.1 Mouse

Infection of inbred strains of mice with *L. major* provides the best model for the immunoregulation that occurs during a cell-mediated response to an intracellular pathogen. In susceptible BALB/c mice, cutaneous *L. major* infection provokes a progressive disease with visceral dissemination which is associated with the appearance in the draining lymph nodes of parasite-specific Th2-type cells, i.e. CD4+ effector T-cells that release IL-4 after stimulation. Conversely, most inbred strains (e.g. CBA, C3H, or C57BL/6) readily control infection and develop a robust response associated with the appearance of Th1-type cells, i.e. CD4+ effector T-cells that release IFN-γ, but not IL-4, and are required for activation of phagocyte-dependent immunity. The capacity of *Leishmania*-specific CD4+ T-cells to passively transfer exacerbation or resistance of disease to naive recipients correlates with their production of Th2 or Th1 cytokines. Then the question is how the Th1 or Th2 commitment is determined. Although some studies have suggested that some leishmanial antigens might drive Th1- or Th2-type responses in susceptible mice (Scott 2003), there is evidence that most naive T-cells can mature into either subset of effector cells and that cytokines supplied at the time of T-cell priming mediate this differentiation. IL-4 (and at less extent IL-10) and IL-12 have been shown to have a major role in mediating Th2 and Th1 development, respectively, whereas the Th1 cytokine IFN-γ apparently does not alter the course of infection in susceptible mice. Early macrophage production of IL-12 suppresses IL-4 transcription, while activating IFN-γ, and persistence of endogenous IL-12 is required to sustain Th1 responses and long-term control of infection. Genetic susceptibility of BALB/c mice to *L. major* probably originates from an aberrant recognition of leishmanial antigen epitopes by CD4+ T-cells, such that IL-4 production becomes overabundant during priming with subsequent loss of IL-12 responsiveness (Locksley and Louis 1992).

Different *Leishmania* species cause clinically distinct diseases and the severity of the disease caused by any given parasite can vary markedly between individual hosts. Till date, two host systems have been classified for studying *Leishmania* infection on the basis of susceptibility and resistance of the host. This observation extends to the murine *L. major* model where the strain of inbred mouse determines the outcome of infection, C57BL/6 mice being uniformly resistant and BALB/c consistently susceptible. It is well documented that Th1 immune response is the key event to prevent *Leishmania* infection. Activated Th1 cells induce IFN-γ that in turns activates the macrophages and kill the parasites. C57BL/6 mice mount early Th1 immune response and prevent the further growth of the parasite causes self-healing phenotype whereas susceptible BALB/c strain mounts early Th2 response and results in non healing lesion and exaggeration of disease. Respective resistance and susceptibility of C57BL/6 and BALB/c strains depend not only on the Th1 and Th2 type of immune response of CD4+ T cells but also on the genetic background of the host. Initially it was shown that resistance or susceptibility of the recombinant strains of mouse was dictated by the haplotype of the host. Congenic mouse of a particular haplotype with either susceptible or resistant background could not correlate the susceptibility or resistance with the haplotype of the strain for the *Leishmania* parasites. It suggests that susceptibility or resistance of the host may be partly regulated by the haplotype with some other factors. Factors for susceptibility or resistance could be segregated by repetitive backcrossing of resistant B10.D2 and susceptible BALB/c strains. Loci on chromosomes were associated with resistance, demonstrating the multigenic nature of this phenotype. Moreover, F1 progeny of BALB/c and C57BL/6 mice were shown to intermediary phenotypes for *Leishmania* infection suggested the contribution of genes either in susceptibility or resistance of the host. Bone marrow macrophages derived under influence of granulocytes macrophage-colony stimulating factor (GM-CSF) or IL-3 or monocytes-colony stimulating factor (M-CSF) further increase the respective resistance and susceptibility of these macrophages to *Leishmania* infection. These observations suggest the critical role of myeloid cells in the resistance or susceptibility to *Leishmania*. Resistance or susceptibility of myeloid cells to *Leishmania* needs to be characterized further. In general, the immune responses following infection of inbred mouse strains with viscerotropic *Leishmania* species, such as *L. donovani* or *L. infantum*, are similar to those observed in the *L. major* mouse model. However, BALB/c mice do not appear to exhibit a similarly high susceptibility to these parasites, since intravenous injection of visceral *Leishmania* results in a self-healing chronic infection. Furthermore, cytokine phenotypes elicited by viscerotropic *Leishmania* in this mouse model are not typical of a Th2-type response.

7.2   Hamster

While the mice are either intrinsically resistant or susceptible to *Leishmania* infection and offer a well-characterized genetic makeup, chiefly by the use of inbred, recombinant and naturally or experimentally mutated strains, hamsters provide an excellent model for an overtly susceptible host. Therefore, hamsters are used for histopathological studies, drug efficiency studies and vaccine studies despite the lack fine immunochemicals that limit the mechanistic exploration of immune responses to *Leishmania* infection. Guinea pigs have been a traditional model for studies of delayed type hypersensitivity. They are the natural host of *L. enriettii* and have been experimentally infected with other species of *Leishmania*. They have been used as a skin-test model to screen potential antigens for use in diagnostic tests for *Leishmania*.

7.3   Dog

In comparison with the bulk of experimental data on murine *Leishmania* models and, at lesser extent, on human leishmaniases, knowledge on the mechanisms involved in the immune response to *Leishmania* in dogs is still limited. This is mainly due to the lack of standardized commercial products for the characterization of canine immune cell populations and chemokines. Recently, DNAsequences of relevant dog cytokines have become available in gene data banks and preliminary evaluations on their use in reverse transcription–PCR assays are in progress. Natural and experimental clinical resistance of dogs to *L. infantum* infection has long been reported. Clinically, three categories of infected asymptomatic dogs can be identi-fied, i.e. those progressing towards overt disease (pre-patent cases), those remaining symptomless for prolonged periods (even for life) and those healing spontaneously. Dogs of the last two groups are usually regarded as resistant, in contrast to animals prone to clinical disease and defined as susceptible. Cross-sectional studies in endemic areas have shown that the ratio between the categories of asymptomatic dogs is approximately 1:1:1, but this depends on the method used to assess the leish-manial infection. Immunologically, this broad spectrumwas confirmed by serologi-cal and peripheral blood mononuclear cells (PBMCs) analysis of a large population of asymptomatic dogs living in an endemic area of Portugal. Results have shown that animals could be equally divided into subjects that had negative antibody response, but a positive T-cell proliferative response to *Leishmania* antigen or posi-tive antibodies but negative T-cell, or were both antibody and T-cell positive. Dog genetics appears to have some influence in determining resistance to clinical leish-maniasis; for instance, among Ibizian hounds, the prevalence of overt disease cases is much lower than in other breeds, and such characteristic has been associated to significant immune cell reactivity as shown by the high rate of positive LST responses. PBMC from both experimentally- and naturally-infected animals show-ing different degrees of clinical expression were examined for their lymphoprolif-erative response to stimulant and *Leishmania* antigen, and for the production of some cytokines, which however did not include IFN-γ or IL-4. In general, lympho-cytes from resistant dogs responded more vigorously to antigen and produced higher levels of IL-2 and TNF than did cells from uninfected or symptomatic dogs. Furthermore, low or undetectable specific antibodies correlated with significant response to LST in asymptomatic dogs, in contrast with high-antibody titers and low-LST response shown in symptomatic animals. Similar results have been reported in subsequent studies, which substantiated the profound suppression of lymphoproliferative response to several stimulants and leishmanial antigens in dogs with signs of leishmaniasis or in asymptomatic dogs progressing to clinical disease. Partial or full restoration of antigen-stimulated lymphocyte blastogenesis was reported in dogs clinically cured following treatment with antimonials or ampho-tericin B. Analysis of PBMC subsets has revealed that such immune-cell restoration was associated with a significant increase in the percentage of CD4+ T-cells. Differential responses of IgG1 and IgG2 subclasses have been proposed as an indi-cator of dichotomous antibody response to *L. infantum* infection, with IgG2 being associated with asymptomatic condition and IgG1 with disease.

References

Abdallah KA, Nour BY, Schallig HD, Mergani A, Hamid Z, Elkarim AA, Saeed OK, Mohamadani AA (2004) Evaluation of the direct agglutination test based on freeze-dried Leishmania donovani promastigotes for the serodiagnosis of visceral leishmaniasis in Sudanese patients. Trop Med Int Health 9:1127–1131

Aebischer T, Moody SF, Handman E (1993) Persistence of virulent Leishmania major in murine cutaneous leishmaniasis: a possible hazard for the host. Infect Immun 61(1):220–226

Aggarwal P, Handa R, Singh S, Wali JP (1999) Kala-azar–new developments in diagnosis and treatment. Indian J Pediatr 66:63–71

Alexander J (1988) Sex differences and cross-immunity in DBA/2 mice infected with L. mexicana and L. major. Parasitology 96(Pt 2):297–302

Alexander J, Bryson K (2005) T helper (h)1/Th2 and Leishmania: paradox rather than paradigm. Immunol Lett 99:17–23

Alvar J, Canavate C, Gutierrez-Solar B, Jimenez M, Laguna F, Lopez-Velez R, Molina R, Moreno J (1997) Leishmania and human immunodeficiency virus coinfection: the first 10 years. Clin Microbiol Rev 10:298–319

Amaral VF, Teva A, Oliveira-Neto MP, Silva AJ, Pereira MS, Cupolillo E, Porrozzi R, Coutinho SG, Pirmez C, Beverley SM, Grimaldi G Jr (2002) Study of the safety, immunogenicity and efficacy of attenuated and killed Leishmania (Leishmania) major vaccines in a rhesus monkey (Macaca mulatta) model of the human disease. Mem Inst Oswaldo Cruz 97:1041–1048

Basu R, Bhaumik S, Basu JM, Naskar K, De T, Roy S (2005) Kinetoplastid membrane protein-11 DNA vaccination induces complete protection against both pentavalent antimonial-sensitive and -resistant strains of Leishmania donovani that correlates with inducible nitric oxide synthase activity and IL-4 generation: evidence for mixed Th1- and Th2-like responses in visceral leishmaniasis. J Immunol 174:7160–7171

Belkaid Y, Mendez S, Lira R, Kadambi N, Milon G, Sacks D (2000) A natural model of Leishmania major infection reveals a prolonged "silent" phase of parasite amplification in the skin before the onset of lesion formation and immunity. J Immunol 165:969–977

Berman JD, Gallalee JV (1987) In vitro antileishmanial activity of inhibitors of steroid biosynthesis and combinations of antileishmanial agents. J Parasitol 73:671–673

Berman JD, Edwards N, King M, Grogl M (1989) Biochemistry of Pentostam resistant Leishmania. Am J Trop Med Hyg 40:159–164

Bernier R, Barbeau B, Tremblay MJ, Olivier M (1998) The lipophosphoglycan of Leishmania donovani up-regulates HIV-1 transcription in T cells through the nuclear factor-kappaB elements. J Immunol 160:2881–2888

Boelaert M, Criel B, Leeuwenburg J, Van Damme W, Le Ray D, Van der Stuyft P (2000) Visceral leishmaniasis control: a public health perspective. Trans R Soc Trop Med Hyg 94:465–471

Bogdan C, Streck H, Rollinghoff M, Solbach W (1989) Cyclosporin A enhances elimination of intracellular L. major parasites by murine macrophages. Clin Exp Immunol 75:141–146

Bogitsh BJ, Middleton OL, Ribeiro-Rodrigues R (1999) Effects of the antitubulin drug trifluralin on the proliferation and metacyclogenesis of Trypanosoma cruzi epimastigotes. Parasitol Res 85:475–480

Bora D (1999) Epidemiology of visceral leishmaniasis in India. Natl Med J India 12:62–68

Bray RS, Lainson R (1966) The immunology and serology of leishmanisis. IV. Results of Ouchterlony double diffusion tests. Trans R Soc Trop Med Hyg 60:605–609

Brochu C, Wang J, Roy G, Messier N, Wang XY, Saravia NG, Ouellette M (2003) Antimony uptake systems in the protozoan parasite Leishmania and accumulation differences in antimony-resistant parasites. Antimicrob Agents Chemother 47:3073–3079

Bryceson AD, Turk JL (1971) The effect of prolonged treatment with antilymphocyte serum on the course of infections with BCG and Leishmania enriettii in the guinea-pig. J Pathol 104:153–165

Bunn-Moreno MM, Madeira ED, Miller K, Menezes JA, Campos-Neto A (1985) Hypergammaglobulinaemia in Leishmania donovani infected hamsters: possible association with a polyclonal activator of B cells and with suppression of T cell function. Clin Exp Immunol 59:427–434

Campos-Neto A, Soong L, Cordova JL, Sant'Angelo D, Skeiky YA, Ruddle NH, Reed SG, Janeway C Jr, McMahon-Pratt D (1995) Cloning and expression of a Leishmania donovani gene instructed by a peptide isolated from major histocompatibility complex class II molecules of infected macrophages. J Exp Med 182:1423–1433

Chang CS, Chang KP (1985) Heme requirement and acquisition by extracellular and intracellular stages of Leishmania mexicana amazonensis. Mol Biochem Parasitol 16:267–276

Chapman WL Jr, Hanson WL, Hendricks LD (1981) Leishmania donovani in the owl monkey (Aotus trivirgatus). Trans R Soc Trop Med Hyg 75:124–125

Clinton BA, Stauber LA, Palczuk NC (1969) Leishmania donovani: antibody response to chicken ovalbumin by infected golden hamsters. Exp Parasitol 25:171–180

Cook JA, Holbrook TW (1983) Immunogenicity of soluble and particulate antigens from Leishmania donovani: effect of glucan as an adjuvant. Infect Immun 40:1038–1043

Croft SL, Engel J (2006) Miltefosine–discovery of the antileishmanial activity of phospholipid derivatives. Trans R Soc Trop Med Hyg 100(Suppl 1):S4–S8

Croft AM, Taylor NA, Rodenhurst KE (2006) Sandflies & leishmaniasis. Lancet 367:112

Cunningham ML, Fairlamb AH (1995) Trypanothione reductase from Leishmania donovani. Purification, characterisation and inhibition by trivalent antimonials. Eur J Biochem 230:460–468

Dasgupta B, Roychoudhury K, Ganguly S, Kumar Sinha P, Vimal S, Das P, Roy S (2003) Antileishmanial drugs cause up-regulation of IFN-γ receptor 1, not only in the monocytes of visceral leishmaniasis cases but also in cultured THP1 cells. Ann Trop Med Parasitol 97:245–257

Desjeux P (1992) Human leishmaniases: epidemiology and public health aspects. World Health Stat Q 45:267–275

Desjeux P (2004) Leishmaniasis. Nat Rev Microbiol 2:692

Desjeux P, Alvar J (2003) Leishmania/HIV co-infections: epidemiology in Europe. Ann Trop Med Parasitol 97(Suppl 1):3–15

Desjeux P, Piot B, O'Neill K, Meert JP (2001) Co-infections of leishmania/HIV in south Europe. Med Trop (Mars) 61:187–193

Dominguez M, Torano A (1999) Immune adherence-mediated opsonophagocytosis: the mechanism of Leishmania infection. J Exp Med 189:25–35

Ephros M, Bitnun A, Shaked P, Waldman E, Zilberstein D (1999) Stage-specific activity of pentavalent antimony against Leishmania donovani axenic amastigotes. Antimicrob Agents Chemother 43:278–282

Evans TG, Smith D, Pearson RD (1990) Humoral factors and nonspecific immune suppression in Syrian hamsters infected with Leishmania donovani. J Parasitol 76:212–217

Fragaki K, Suffia I, Ferrua B, Rousseau D, Le Fichoux Y, Kubar J (2001) Immunisation with DNA encoding Leishmania infantum protein papLe22 decreases the frequency of parasitemic episodes in infected hamsters. Vaccine 19:1701–1709

Ganguly A, Das BB, Sen N, Roy A, Dasgupta SB, Majumder HK (2006) 'LeishMan' topoisomerase I: an ideal chimera for unraveling the role of the small subunit of unusual bi-subunit topoisomerase I from Leishmania donovani. Nucleic Acids Res 34:6286–6297

Garg R, Dube A (2006) Animal models for vaccine studies for visceral leishmaniasis. Indian J Med Res 123:439–454

Ghalib HW, Piuvezam MR, Skeiky YA, Siddig M, Hashim FA, el-Hassan AM, Russo DM, Reed SG (1993) Interleukin 10 production correlates with pathology in human Leishmania donovani infections. J Clin Invest 92:324–329

Ghose AC, Haldar JP, Pal SC, Mishra BP, Mishra KK (1980) Serological investigations on Indian kala-azar. Clin Exp Immunol 40:318–326

Ghosh MK, Nandy A, Addy M, Maitra TK, Ghose AC (1995) Subpopulations of T lymphocytes in the peripheral blood, dermal lesions and lymph nodes of post kala-azar dermal leishmaniasis patients. Scand J Immunol 41:11–17

Gicheru MM, Olobo JO, Anjili CO, Orago AS, Modabber F, Scott P (2001) Vervet monkeys vaccinated with killed Leishmania major parasites and interleukin-12 develop a type 1 immune response but are not protected against challenge infection. Infect Immun 69:245–251

Glaser TA, Baatz JE, Kreishman GP, Mukkada AJ (1988) pH homeostasis in Leishmania donovani amastigotes and promastigotes. Proc Natl Acad Sci USA 85:7602–7606

Gomes RF, Macedo AM, Pena SD, Melo MN (1995) Leishmania (Viannia) braziliensis: genetic relationships between strains isolated from different areas of Brazil as revealed by DNA fingerprinting and RAPD. Exp Parasitol 80:681–687

Goto H, Lindoso JA (2004) Immunity and immunosuppression in experimental visceral leishmaniasis. Braz J Med Biol Res 37:615–623

Goyard S, Segawa H, Gordon J, Showalter M, Duncan R, Turco SJ, Beverley SM (2003) An in vitro system for developmental and genetic studies of Leishmania donovani phosphoglycans. Mol Biochem Parasitol 130:31–42

Grondin K, Haimeur A, Mukhopadhyay R, Rosen BP, Ouellette M (1997) Co-amplification of the gamma-glutamylcysteine synthetase gene gsh1 and of the ABC transporter gene pgpA in arsenite-resistant Leishmania tarentolae. EMBO J 16:3057–3065

Guerin PJ, Olliaro P, Sundar S, Boelaert M, Croft SL, Desjeux P, Wasunna MK, Bryceson AD (2002) Visceral leishmaniasis: current status of control, diagnosis, and treatment, and a proposed research and development agenda. Lancet Infect Dis 2:494–501

Haimeur A, Guimond C, Pilote S, Mukhopadhyay R, Rosen BP, Poulin R, Ouellette M (1999) Elevated levels of polyamines and trypanothione resulting from overexpression of the ornithine decarboxylase gene in arsenite-resistant Leishmania. Mol Microbiol 34:726–735

Handman E (1997) Leishmania vaccines: old and new. Parasitol Today 13:236–238

Handman E (1999) Cell biology of Leishmania. Adv Parasitol 44:1–39

Handman E (2001) Leishmaniasis: current status of vaccine development. Clin Microbiol Rev 14:229–243

Handman E, Jarvis HM (1985) Nitrocellulose-based assays for the detection of glycolipids and other antigens: mechanism of binding to nitrocellulose. J Immunol Methods 83:113–123

Handman E, Ceredig R, Mitchell GF (1979) Murine cutaneous leishmaniasis: disease patterns in intact and nude mice of various genotypes and examination of some differences between normal and infected macrophages. Aust J Exp Biol Med Sci 57:9–29

Handman E, Symons FM, Baldwin TM, Curtis JM, Scheerlinck JP (1995) Protective vaccination with promastigote surface antigen 2 from Leishmania major is mediated by a TH1 type of immune response. Infect Immun 63:4261–4267

Heinzel FP, Sadick MD, Mutha SS, Locksley RM (1991) Production of interferon gamma, interleukin 2, interleukin 4, and interleukin 10 by CD4+ lymphocytes in vivo during healing and progressive murine leishmaniasis. Proc Natl Acad Sci USA 88:7011–7015

Herwaldt BL (1999) Leishmaniasis. Lancet 354:1191–1199

Hill JO (1987) Modulation of the pattern of development of experimental disseminated leishmaniasis by Corynebacterium parvum. J Leukoc Biol 41:165–169

Holaday BJ, Sadick MD, Wang ZE, Reiner SL, Heinzel FP, Parslow TG, Locksley RM (1991)
Reconstitution of Leishmania immunity in severe combined immunodeficient mice using Th1-
and Th2-like cell lines. J Immunol 147:1653–1658

Howard JG, Liew FY (1984) Mechanisms of acquired immunity in leishmaniasis. Philos Trans R
Soc Lond B Biol Sci 307:87–98

Howard JG, Hale C, Chan-Liew WL (1980) Immunological regulation of experimental cutaneous
leishmaniasis. 1. Immunogenetic aspects of susceptibility to Leishmania tropica in mice.
Parasite Immunol 2:303–314

Howard JG, Liew FY, Hale C, Nicklin S (1984) Prophylactic immunization against experimental
leishmaniasis. II. Further characterization of the protective immunity against fatal Leishmania
tropica infection induced by irradiated promastigotes. J Immunol 132:450–455

Islam MZ, Itoh M, Mirza R, Ahmed I, Ekram AR, Sarder AH, Shamsuzzaman SM, Hashiguchi Y,
Kimura E (2004) Direct agglutination test with urine samples for the diagnosis of visceral
leishmaniasis. Am J Trop Med Hyg 70:78–82

Jaffe CL, McMahon-Pratt D (1987) Serodiagnostic assay for visceral leishmaniasis employing
monoclonal antibodies. Trans R Soc Trop Med Hyg 81:587–594

Jaffe CL, Rachamim N, Sarfstein R (1990) Characterization of two proteins from Leishmania
donovani and their use for vaccination against visceral leishmaniasis. J Immunol 144:699–706

Jensen AT, Curtis J, Montgomery J, Handman E, Theander TG (2001) Molecular and immunological characterisation of the glucose regulated protein 78 of Leishmania donovani(1). Biochim
Biophys Acta 1549:73–87

Jha TK (1983) Evaluation of allopurinol in the treatment of kala-azar occurring in North Bihar,
India. Trans R Soc Trop Med Hyg 77:204–207

Kamhawi S (2000) The biological and immunomodulatory properties of sand fly saliva and its role
in the establishment of Leishmania infections. Microbes Infect 2:1765–1773

Karp CL, el-Safi SH, Wynn TA, Satti MM, Kordofani AM, Hashim FA, Hag-Ali M, Neva FA,
Nutman TB, Sacks DL (1993) In vivo cytokine profiles in patients with kala-azar. Marked
elevation of both interleukin-10 and interferon-gamma. J Clin Invest 91:1644–1648

Kaye PM, Svensson M, Ato M, Maroof A, Polley R, Stager S, Zubairi S, Engwerda CR (2004) The
immunopathology of experimental visceral leishmaniasis. Immunol Rev 201:239–253

Kelley JM (1997) Genetic transformation of parasitic protozoa. Adv Parasitol 39:227–270

Kemp M, Kurtzhals JA, Bendtzen K, Poulsen LK, Hansen MB, Koech DK, Kharazmi A, Theander
TG (1993) Leishmania donovani-reactive Th1- and Th2-like T-cell clones from individuals
who have recovered from visceral leishmaniasis. Infect Immun 61:1069–1073

Kemp M, Handman E, Kemp K, Ismail A, Mustafa MD, Kordofani AY, Bendtzen K, Kharazmi A,
Theander TG (1998) The Leishmania promastigote surface antigen-2 (PSA-2) is specifically
recognised by Th1 cells in humans with naturally acquired immunity to L. major. FEMS
Immunol Med Microbiol 20:209–218

Kemp K, Kemp M, Kharazmi A, Ismail A, Kurtzhals JA, Hviid L, Theander TG (1999) Leishmania-
specific T cells expressing interferon-gamma (IFN-gamma) and IL-10 upon activation are
expanded in individuals cured of visceral leishmaniasis. Clin Exp Immunol 116:500–504

Kima PE, Soong L, Chicharro C, Ruddle NH, McMahon-Pratt D (1996) Leishmania-infected macrophages sequester endogenously synthesized parasite antigens from presentation to CD4+ T
cells. Eur J Immunol 26(12):3163–3169

Kishore K, Kumar V, Kesari S, Dinesh DS, Kumar AJ, Das P, Bhattacharya SK (2006) Vector
control in leishmaniasis. Indian J Med Res 123:467–472

Lehmann J, Enssle KH, Lehmann I, Emmendörfer A, Lohmann-Matthes ML (2000) The capacity
to produce IFN-gamma rather than the presence of interleukin-4 determines the resistance and
the degree of susceptibility to Leishmania donovani infection in mice. J Interferon Cytokine
Res 20(1):63–77

Lehn M, Kandil O, Arena C, Rein MS, Remold HG (1992) Interleukin-4 fails to inhibit interferon-gamma-induced activation of human colostral macrophages. Cell Immunol 141:233–242

Leifso K, Cohen-Freue G, Dogra N, Murray A, McMaster WR (2007) Genomic and proteomic expression analysis of Leishmania promastigote and amastigote life stages: the Leishmania genome is constitutively expressed. Mol Biochem Parasitol 152:35–46

Liew FY, O'Donnell CA (1993) Immunology of leishmaniasis. Adv Parasitol 32:161–259

Liu Z, Shen J, Carbrey JM, Mukhopadhyay R, Agre P, Rosen BP (2002) Arsenite transport by mammalian aquaglyceroporins AQP7 and AQP9. Proc Natl Acad Sci USA 99(9):6053–6058

Locksley RM, Louis JA (1992) Immunology of leishmaniasis. Curr Opin Immunol 4:413–418

Manson-Bahr PE (1971) Leishmaniasis. Int Rev Trop Med 4:123–140

Mayrink W, da Costa CA, Magalhaes PA, Melo MN, Dias M, Lima AO, Michalick MS, Williams P (1979) A field trial of a vaccine against American dermal leishmaniasis. Trans R Soc Trop Med Hyg 73:385–387

Mayrink W, Williams P, da Costa CA, Magalhaes PA, Melo MN, Dias M, Oliveira Lima A, Michalick MS, Ferreira Carvalho E, Barros GC et al (1985) An experimental vaccine against American dermal leishmaniasis: experience in the State of Espirito Santo, Brazil. Ann Trop Med Parasitol 79:259–269

McConville MJ, Ralton JE (1997) Developmentally regulated changes in the cell surface architecture of Leishmania parasites. Behring Inst Mitt Mar(99):34–43

Medzihradszky KF, Campbell JM, Baldwin MA, Falick AM, Juhasz P, Vestal ML, Burlingame AL (2000) The characteristics of peptide collision-induced dissociation using a high-performance MALDI-TOF/TOF tandem mass spectrometer. Anal Chem 72:552–558

Melby PC, Chandrasekar B, Zhao W, Coe JE (2001) The hamster as a model of human visceral leishmaniasis: progressive disease and impaired generation of nitric oxide in the face of a prominent Th1-like cytokine response. J Immunol 166:1912–1920

Mitchell GF, Handman E, Spithill TW (1984) Vaccination against cutaneous leishmaniasis in mice using nonpathogenic cloned promastigotes of Leishmania major and importance of route of injection. Aust J Exp Biol Med Sci 62(Pt 2):145–153

Modabber F (1990) Development of vaccines against leishmaniasis. Scand J Infect Dis Suppl 76:72–78

Modabber F (1995) Vaccines against leishmaniasis. Ann Trop Med Parasitol 89(Suppl 1):83–88

Moll H, Flohe S, Rollinghoff M (1995) Dendritic cells in Leishmania major-immune mice harbor persistent parasites and mediate an antigen-specific T cell immune response. Eur J Immunol 25:693–699

Mottram JC, Coombs GH (1998) Leishmania cysteine proteinases: virulence factors in quest of a function-reply. Parasitol Today 14:251–252

Mukhopadhyay R, Dey S, Xu N, Gage D, Lightbody J, Ouellette M, Rosen BP (1996) Trypanothione overproduction and resistance to antimonials and arsenicals in Leishmania. Proc Natl Acad Sci USA 93:10383–10387

Murray HW (1982) Cell-mediated immune response in experimental visceral leishmaniasis. II. Oxygen-dependent killing of intracellular Leishmania donovani amastigotes. J Immunol 129:351–357

Murray HW (1997) Endogenous interleukin-12 regulates acquired resistance in experimental visceral leishmaniasis. J Infect Dis 175(6):1477–1479

Murray HW (2005) Prevention of relapse after chemotherapy in a chronic intracellular infection: mechanisms in experimental visceral leishmaniasis. J Immunol 174:4916–4923

Murray HW, Rubin BY, Rothermel CD (1983) Killing of intracellular Leishmania donovani by lymphokine-stimulated human mononuclear phagocytes. Evidence that interferon-gamma is the activating lymphokine. J Clin Invest 72:1506–1510

Murray HW, Montelibano C, Peterson R, Sypek JP (2000) Interleukin-12 regulates the response to chemotherapy in experimental visceral Leishmaniasis. J Infect Dis 182:1497–1502

Murray HW, Berman JD, Davies CR, Saravia NG (2005) Advances in leishmaniasis. Lancet 366:1561–1577

Ouellette M, Borst P (1991) Drug resistance and P-glycoprotein gene amplification in the proto-zoan parasite Leishmania. Res Microbiol 142:737–746

Pal R, Anuradha, Rizvi SY, Kundu B, Mathur KB, Katiyar JC (1991) Leishmania donovani in hamsters: stimulation of non-specific resistance by some novel glycopeptides and impact on therapeutic efficacy. Experientia 47:486–490

Passos S, Carvalho LP, Orge G, Jeronimo SM, Bezerra G, Soto M, Alonso C, Carvalho EM (2005) Recombinant leishmania antigens for serodiagnosis of visceral leishmaniasis. Clin Diagn Lab Immunol 12:1164–1167

Passwell JH, Shor R, Smolen J, Jaffe CL (1994) Infection of human monocytes by Leishmania results in a defective oxidative burst. Int J Exp Pathol 75(4):277–84

Perez-Victoria JM, Di Pietro A, Barron D, Ravelo AG, Castanys S, Gamarro F (2002) Multidrug resistance phenotype mediated by the P-glycoprotein-like transporter in Leishmania: a search for reversal agents. Curr Drug Targets 3:311–333

Peters W, Evans DA, Lanham SM (1983) Importance of parasite identification in cases of leish-maniasis. J R Soc Med 76:540–542

Pfeffer K, Schoel B, Gulle H, Kaufmann SHE, Wagner H (1990) Primary responses of human T cells to mycobacteria: a frequent set of γδ T cells are stimulated by protease-resistant ligands. Eur J Immunol 20:1175–1179

Prasad LS (1999) Kala azar. Indian J Pediatr 66:539–546

Probst P, Skeiky YA, Steeves M, Gervassi A, Grabstein KH, Reed SG (1997) A Leishmania protein that modulates interleukin (IL)-12, IL-10 and tumor necrosis factor-alpha production and expression of B7-1 in human monocyte-derived antigen-presenting cells. Eur J Immunol 27:2634–2642

Rachamim N, Jaffe CL (1993) Pure protein from Leishmania donovani protects mice against both cutaneous and visceral leishmaniasis. J Immunol 150(6):2322–31

Ramos H, Valdivieso E, Gamargo M, Dagger F, Cohen BE (1996) Amphotericin B kills unicellular leishmanias by forming aqueous pores permeable to small cations and anions. J Membr Biol 152:65–75

Ravindran R, Ali N (2004) Progress in vaccine research and possible effector mechanisms in vis-ceral leishmaniasis. Curr Mol Med 4:697–709

Reed MJ, Purohit A (2001) Aromatase regulation and breast cancer. Clin Endocrinol 54:563–571

Reed SG, Badaro R, Lloyd RM (1987) Identification of specific and cross-reactive antigens of Leishmania donovani chagasi by human infection sera. J Immunol 138:1596–1601

Rees PH, Kager PA, Wellde BT, Hockmeyer WT (1984) The response of Kenyan kala-azar to treat-ment with sodium stibogluconate. Am J Trop Med Hyg 33:357–361

Reiner SL, Locksley RM (1995) The regulation of immunity to Leishmania major. Annu Rev Immunol 13:151–177

Rezai HR, Ardehali SM, Amirhakimi G, Kharazmi A (1978) Immunological features of kala-azar. Am J Trop Med Hyg 27:1079–1083

Rittig MG, Schroppel K, Seack KH, Sander U, N'Diaye EN, Maridonneau-Parini I, Solbach W, Bogdan C (1998) Coiling phagocytosis of trypanosomatids and fungal cells. Infect Immun 66:4331–4339

Roberts WL, Hariprashad J, Rainey PM, Murray HW (1996) Pentavalent antimony-mannan con-jugate therapy of experimental visceral leishmaniasis. Am J Trop Med Hyg 55:444–446

Rodrigues V Jr, Santana da Silva J, Campos-Neto A (1998) Transforming growth factor beta and immunosuppression in experimental visceral leishmaniasis. Infect Immun 66:1233–1236

Russell DG, Alexander J (1988) Effective immunization against cutaneous leishmaniasis with defined membrane antigens reconstituted into liposomes. J Immunol 140:1274–1279

Ryan KA, Garraway LA, Descoteaux A, Turco SJ, Beverley SM (1993) Isolation of virulence genes directing surface glycosyl-phosphatidylinositol synthesis by functional complementa-tion of Leishmania. Proc Natl Acad Sci USA 90:8609–8613

Sacks DL, Lal SL, Shrivastava SN, Blackwell J, Neva FA (1987) An analysis of T cell responsive-ness in Indian kala-azar. J Immunol 138:908–913

Sacks DL, Pimenta PF, McConville MJ, Schneider P, Turco SJ (1995) Stage-specific binding of Leishmania donovani to the sand fly vector midgut is regulated by conformational changes in the abundant surface lipophosphoglycan. J Exp Med 181:685–697

Saha S, Mondal S, Banerjee A, Ghose J, Bhowmick S, Ali N (2006) Immune responses in kala-azar. Indian J Med Res 123:245–266

Salotra P, Singh R (2006) Challenges in the diagnosis of post kala-azar dermal leishmaniasis. Indian J Med Res 123:295–310

Santos WR, de Lima VM, de Souza EP, Bernardo RR, Palatnik M, Palatnik de Sousa CB (2002) Saponins, IL12 and BCG adjuvant in the FML-vaccine formulation against murine visceral leishmaniasis. Vaccine 21:30–43

Saravia NG, Hazbon MH, Osorio Y, Valderrama L, Walker J, Santrich C, Cortazar T, Lebowitz JH, Travi BL (2005) Protective immunogenicity of the paraflagellar rod protein 2 of Leishmania mexicana. Vaccine 23:984–995

Schlein Y, Jacobson RL, Messer G (1992) Leishmania infections damage the feeding mechanism of the sandfly vector and implement parasite transmission by bite. Proc Natl Acad Sci USA 89:9944–9948

Scott P (2003) Development and regulation of cell-mediated immunity in experimental leishmaniasis. Immunol Res 27:489–498

Scott P (2005) Immunologic memory in cutaneous leishmaniasis. Cell Microbiol 7:1707–1713

Scott DA, Coombs GH, Sanderson BE (1987) Effects of methotrexate and other antifolates on the growth and dihydrofolate reductase activity of Leishmania promastigotes. Biochem Pharmacol 36:2043–2045

Scott P, Natovitz P, Coffman RL, Pearce E, Sher A (1988) Immunoregulation of cutaneous leishmaniasis. T cell lines that transfer protective immunity or exacerbation belong to different T helper subsets and respond to distinct parasite antigens. J Exp Med 168:1675–1684

SenGupta PC (1962) Pathogenicity of Leishmania donovani in man. Rev Inst Med Trop Sao Paulo 4:130–135

Sereno D, Holzmuller P, Mangot I, Cuny G, Ouaissi A, Lemesre JL (2001) Antimonial-mediated DNA fragmentation in Leishmania infantum amastigotes. Antimicrob Agents Chemother 45:2064–2069

Shaked-Mishan P, Ulrich N, Ephros M, Zilberstein D (2001) Novel Intracellular SbV reducing activity correlates with antimony susceptibility in Leishmania donovani. J Biol Chem 276:3971–3976

Singh N (2006) Drug resistance mechanisms in clinical isolates of Leishmania donovani. Indian J Med Res 123:411–422

Sinha J, Raay B, Das N, Medda S, Garai S, Basu MK (2002) Bacopasaponin C: critical evaluation of anti-leishmanial properties in various delivery modes. Drug Deliv 9:55–62

Sinha PK, Bimal S, Singh SK, Pandey K, Gangopadhyay DN, Bhattacharya SK (2006) Pre- & post-treatment evaluation of immunological features in Indian visceral leishmaniasis (VL) patients with HIV co-infection. Indian J Med Res 123:197–202

Solbach W, Laskay T (2000) The host response to Leishmania infection. Adv Immunol 74:275–317

Soto M, Requena JM, Quijada L, Perez MJ, Nieto CG, Guzman F, Patarroyo ME, Alonso C (1999) Antigenicity of the Leishmania infantum histones H2B and H4 during canine viscerocutaneous leishmaniasis. Clin Exp Immunol 115:342–349

Squires KE, Schreiber RD, McElrath MJ, Rubin BY, Anderson SL, Murray HW (1989) Experimental visceral leishmaniasis: role of endogenous IFN-gamma in host defense and tissue granulomatous response. J Immunol 143(12):4244–4249

Stauber LA (1963) Immunity to leishmania. Ann NY Acad Sci 113:409–417

Sternberg J, Turner CM, Wells JM, Ranford-Cartwright LC, Le Page RW, Tait A (1989) Gene exchange in African trypanosomes: frequency and allelic segregation. Mol Biochem Parasitol 34:269–279

Stobie L, Gurunathan S, Prussin C, Sacks DL, Glaichenhaus N, Wu CY, Seder RA (2000) The role of antigen and IL-12 in sustaining Th1 memory cells in vivo: IL-12 is required to maintain memory/effector Th1 cells sufficient to mediate protection to an infectious parasite challenge. Proc Natl Acad Sci USA 97:8427–8432

Suffia I, Ferrua B, Stien X, Mograbi B, Marty P, Rousseau D, Fragaki K, Kubar J (2000) A novel Leishmania infantum recombinant antigen which elicits interleukin 10 production by peripheral blood mononuclear cells of patients with visceral leishmaniasis. Infect Immun 68:630–636

Sukumaran B, Tewary P, Saxena S, Madhubala R (2003) Vaccination with DNA encoding ORFF antigen confers protective immunity in mice infected with Leishmania donovani. Vaccine 21:1292–1299

Sundar S (2001a) Drug resistance in Indian visceral leishmaniasis. Trop Med Int Health 6:849–854

Sundar S (2001b) Liposomal amphotericin B. Lancet 357:801–802

Sundar S (2003) Diagnosis of kala-azar–an important stride. J Assoc Phys India 51:753–755

Sundar S, Chatterjee M (2006) Visceral leishmaniasis - current therapeutic modalities. Indian J Med Res 123:345–352

Sundar S, Rai M (2002) Advances in the treatment of leishmaniasis. Curr Opin Infect Dis 15:593–598

Sundar S, Reed SG, Singh VP, Kumar PC, Murray HW (1998) Rapid accurate field diagnosis of Indian visceral leishmaniasis. Lancet 351:563–565

Sundar S, Jha TK, Thakur CP, Engel J, Sindermann H, Fischer C, Junge K, Bryceson A, Berman J (2002) Oral miltefosine for Indian visceral leishmaniasis. N Engl J Med 347:1739–1746

Sundar S, Agrawal S, Pai K, Chance M, Hommel M (2005) Detection of leishmanial antigen in the urine of patients with visceral leishmaniasis by a latex agglutination test. Am J Trop Med Hyg 73:269–271

Sundar S, Maurya R, Singh RK, Bharti K, Chakravarty J, Parekh A, Rai M, Kumar K, Murray HW (2006) Rapid, noninvasive diagnosis of VL in India: comparison of two immunochromatographic strip tests for detection of anti-K39 antibody. J Clin Microbiol 44:251–253

Sundar S, Jha TK, Thakur CP, Sinha PK, Bhattacharya SK (2007) Injectable paromomycin for Visceral leishmaniasis in India. N Engl J Med 356:2571–2581

Swaminath CS, Shortt HE, Anderson LA (2006) Transmission of Indian kala-azar to man by the bites of Phlebotomus argentipes, ann and brun. 1942. Indian J Med Res 123:473–477

TDR News (2005) TDR communication. WHO, Geneva

Tewary P, Jain M, Sahani MH, Saxena S, Madhubala R (2005) A heterologous prime-boost vaccination regimen using ORFF DNA and recombinant ORFF protein confers protective immunity against experimental visceral leishmaniasis. J Infect Dis 191:2130–2137

Thakur BB (2003) Breakthrough in the management of visceral leishmaniasis. J Assoc Phys India 51:649–651

Thakur CP, Sinha GP, Pandey AK, Barat D, Sinha PK (1993) Amphotericin B in resistant kala-azar in Bihar. Natl Med J India 6:57–60

Thakur CP, Pandey AK, Sinha GP, Roy S, Behbehani K, Olliaro P (1996) Comparison of three treatment regimens with liposomal amphotericin B (AmBisome) for visceral leishmaniasis in India: a randomized dose-finding study. Trans R Soc Trop Med Hyg 90:319–322

Tonui WK, Ngumbi PM, Mpoke SS, Orago AS, Mbati PA, Turco SJ, Mkoji GM (2004) Leishmania major-Phlebotomus duboscqi interactions: inhibition of anti-LPG antibodies and characterisation of two proteins with shared epitopes. East Afr Med J 81:97–103

Valenzuela JG, Belkaid Y, Garfield MK, Mendez S, Kamhawi S, Rowton ED, Sacks DL, Ribeiro JM (2001) Toward a defined anti-Leishmania vaccine targeting vector antigens: characterization of a protective salivary protein. J Exp Med 194:331–342

Vickers TJ, Wyllie S, Fairlamb AH (2004) Leishmania major elongation factor 1B complex has trypanothione S-transferase and peroxidase activity. J Biol Chem 279:49003–49009

Waitumbi JN, Murphy NB (1993) Inter- and intra-species differentiation of trypanosomes by genomic fingerprinting with arbitrary primers. Mol Biochem Parasitol 58:181–185

Weinheber N, Wolfram M, Harbecke D, Aebischer T (1988) Phagocytosis of Leishmania mexicana amastigotes by macrophages leads to a sustained suppression of IL-12 production. Eur J Immunol 28(8):2467–2477

White AC Jr, McMahon-Pratt D (1990) Prophylactic immunization against experimental Leishmania donovani infection by use of a purified protein vaccine. J Infect Dis 161:1313–1314

WHO (2001) Control of leishmaniasis. World Health Organization, Geneva

Wilson ME, Jeronimo SM, Pearson RD (2005) Immunopathogenesis of infection with the visceralizing Leishmania species. Microb Pathog 38:147–160

Wyllie S, Cunningham ML, Fairlamb AH (2004) Dual action of antimonial drugs on thiol redox metabolism in the human pathogen Leishmania donovani. J Biol Chem 279:39925–39932

Yan S, Li F, Ding K, Sun H (2003) Reduction of pentavalent antimony by trypanothione and formation of a binary and ternary complex of antimony(III) and trypanothione. J Biol Inorg Chem 8:689–697

Zehra K, Pal R, Anuradha, Rizvi SY, Haq W, Kundu B, Katiyar JC, Mathur KB (1995) Leishmania donovani in hamsters: stimulation of non-specific resistance by novel lipopeptides and their effect in antileishmanial therapy. Experientia 51:725–730

Zhang T, Maekawa Y, Sakai T, Nakano Y, Ishii K, Hisaeda H, Dainichi T, Asao T, Katunuma N, Himeno K (2001) Treatment with cathepsin L inhibitor potentiates Th2-type immune response in Leishmania major-infected BALB/c mice. Int Immunol 13(8):975–982

Zhang WW, Miranda-Verastegui C, Arevalo J, Ndao M, Ward B, Llanos-Cuentas A, Matlashewski G (2006) Development of a genetic assay to distinguish between Leishmania viannia species on the basis of isoenzyme differences. Clin Infect Dis 42:801–809

Zijlstra EE, el-Hassan AM, Ismael A (1995) Endemic kala-azar in eastern Sudan: post-kala-azar dermal leishmaniasis. Am J Trop Med Hyg 52:299–305